Herefordshire's
Rocks & Scenery
A Geology of the County

Herefordshire's
Rocks & Scenery
A Geology of the County

edited by
John Payne

Logaston Press

LOGASTON PRESS
Little Logaston Woonton Almeley
Herefordshire HR3 6QH
logastonpress.co.uk

First published by Logaston Press 2017
Reprinted 2018
Copyright text © Authors of each chapter
Copyright illustrations © as per credits and acknowledgements

All rights reserved. No part of this publication
may be reproduced, stored in a retrieval system,
or transmitted, in any form or by any means,
electronic, mechanical, photocopying, recording
or otherwise, without the prior permission,
in writing, of the publisher

ISBN 978 1 9010839 16 4

Typeset by Logaston Press
and printed and bound in Poland by
www.lfbookservices.co.uk

Front Cover:
Main picture: The Olchon Valley in west Herefordshire, from the south. (Figure 7.1)
Small pictures, from the top:
Distorted bedding in the Old Red Sandstone on the Cat's Back. (Figure 7.14)
Granite from the Malvern Hills. (Figure 4.8a)
A rugose coral from the Silurian rocks.
Conglomerate rock from the Forest of Dean. (Figure 7.18)
A mass of crinoid fragments from the Silurian limestone.
Pegmatite from the Malvern Hills. (Figure 4.8b)

Contents

	Acknowledgements	vi
	Dedication to Peter Thomson	vii
	Foreword *by* Lawrence Banks	ix
	Preface	xi
	Contributors	xv
	Publications from the Earth Heritage Trust and the British Geological Survey	xix
Chapter 1	An Introduction to Herefordshire *by* Robert Williams	1
Chapter 2	Views from Five Hills *by* Robert Williams (with contributions from Rosamund Skelton and Geoff Steel)	7
Chapter 3	Basic Geology *by* Charles Hopkinson (with contributions from Paul Olver and Robert Williams)	33
Chapter 4	The Precambrian Era – Herefordshire's Foundation *by* Paul Olver	55
Chapter 5	The Cambrian and Ordovician Periods *by* Paul Olver	81
Chapter 6	The Marine Silurian *by* Dave Green	91
Chapter 7	The Old Red Sandstone – all of Herefordshire above sea level *by* Paul Olver	123
Chapter 8	The Carboniferous, Permian and Triassic Periods *by* Dave Green	145
Chapter 9	The Cainozoic Era – Before and After the Ice *by* Moira Jenkins (with contribution from Dave Green)	165
Chapter 10	The Impact of Geology on Herefordshire and its People *by* Charles Hopkinson (with contributions from Kate Andrew and Robert Williams)	195
Appendix	The 'Hereford Speech' *by* John Masefield	209
	Glossary	211
	References	217
	Index	229

Acknowledgements

The editor and authors wish to acknowledge the late Mrs Stephanie Thomson for permission to use the block diagrams produced by her late husband, Peter Thomson, and also Gerry Calderbank for enhancing and up-dating the block diagrams for this publication, as well as for all his work in producing drawings to accompany the text. We also wish to thank the Trustees of the British Museum for permission to use the drawing of Roderick Murchison (Fig. 2.11); Prof. Derek Siveter for permission to use Figure 6.21; the Woolhope Club for permission to use Figure 2.27; and The Geologists' Association for permission to use illustrations adapted from diagrams in the *Proceedings of the Geologists' Association* as Figures 4.6, 4.10 and 4.16. Thanks are due also to the Herefordshire and Worcestershire Earth Heritage Trust for permission to use many illustrations from their publications.

Robert Williams wishes to thank Andrew and Gill Jenkinson of Scenesetters, Welshpool, for kind support and advice over many years – including the concept of 'An Eye for the Land', which formed the theme of a distinctive and memorable summer school held at Lucton in 2010: *An Eye for the Land: An Introduction to Field Geology in the Welsh Marches*, and also David Prentice of Malvern for permission to use his picture *English Air – South Prospect* (Fig. 2.1).

Moira Jenkins wishes to acknowledge the large amount of factual information supplied by Dr Andrew E. Richards, who has done so much work researching the Pleistocene deposits found in Herefordshire.

Many other acknowledgements of assistance need to be made by the editor, both personally and on behalf of the Woolhope Club Geology Section and the main Club itself. The editor wishes to thank all the authors and contributors for the various chapters for their patience, good humour and ready acquiescence when he required modifications or additions to their work. They are Robert Williams, Charles Hopkinson, Paul Olver, Dave Green, Kate Andrew and Moira Jenkins. The other members of the committee of the Geology Section of the Club have always been most encouraging. These include in particular the members of the Publications Working Party (Paul Olver, Charles Hopkinson, Gerry Calderbank, Robert Williams and Mike Rosenbaum) who did much work on the book material prior to the involvement of the present editor. The editor wishes to thank his wife, Valerie, for her strong support during this work. The reviewers and proof readers of the book also merit my grateful appreciation, particularly in those thankfully very few cases where they have pointed out errors in the geology in the text. On behalf of the Geology Section of the Club, I must offer further thanks to the authors as well as to John Stocks, who took many of the photographs, and, especially, Gerry Calderbank, who has generated most of the drawn diagrams within the book. Dave Green provided the inspiration for the palaeogeographic scenes in each chapter. Derek Foxton was our photographer for twenty-four aerial photographs and some other pictures. Finally, both the Section and the main Club are most grateful to the Smith Fund of the Club for financial support and to Lawrence Banks for his support, both financial and in writing the Foreword to this book.

*Dedicated to the memory
of
Peter Thomson (1925-2005)*

The dedication of this book to Peter Thomson's memory arises from his inspirational contributions over many years to the work – and especially the geological work – of the Woolhope Naturalists' Field Club. Peter was a graduate of University College, London. He joined the Woolhope Club in 1963 and from 1999-2000 was the Club's President. To mark the 150th anniversary of the Club's foundation in 1851, he was one of the contributors to *A Herefordshire Miscellany* published by Lapridge Publications in 2000. Over the years, he was county recorder for the Club in both geology and botany – the latter an involvement and interest he shared with his wife, Stephanie. In 2004, Peter was one of a group of geological enthusiasts which set up the current Geology Section of the Woolhope Club; and this book is the result of work done since then by members of the Geology Section.

One of the reasons for setting up the Geology Section was to restore something of the emphasis on Geology in the Woolhope Club that the subject had in the Club's early days (there now being Archaeology and Natural History sections as well). As may be seen in some of the chapters of this book, early members of the Club had a keen interest in the new science of geology – including two of the Club's early Presidents, Revd T.T. Lewis and R.W. Banks (see the Foreword and Chapter 2 of this book). Indeed, in those early years the Club had Honorary Members of national geological importance including Adam Sedgwick, Roderick Murchison and Charles Lyell. Peter's work with others in setting up the Geology Section has thus enabled the Club, once again, to explore matters of local and national geological interest.

Peter had a great gift for sharing his geological scholarship in talks and field visits. Originally trained as a geographer, he was professionally involved in teacher training until his retirement in 1983, after which he devoted his life to natural history, especially geology and botany, in the Marches area. Indeed, as mentioned in his obituary in the Woolhope Club's 2005 *Transactions*, Peter's 'simple explanation of detail added enormously to the enjoyment of his hearers and was much enhanced by his beautiful slides'. As part of his presentations he also produced outstanding block diagrams to illustrate the geology and landscape of key locations in the county and beyond. (See up-graded examples of these in Chapter 2 plus a copy of the original version of 'Dinmore Hill' [below].) At this time he was also instrumental with others in establishing the Herefordshire & Worcestershire RIGS Group, now known as the Herefordshire & Worcestershire Earth Heritage Trust (EHT). For this he contributed to the first EHT Trail Guide, *Wigmore Glacial Lake*, which won a national award for an innovative geological publication, and he was the principal author of the later EHT guide to the *Woolhope Dome*. He published his work on local geomorphology, as well as botany, in the Woolhope Club *Transactions*.

In conclusion, therefore, it is highly appropriate that this book should be dedicated to Peter. Like others, he had come to recognise the possibilities of such a volume and, as a result of so much of his work, there were members of the Woolhope Club inspired to contribute to this. We trust that this book does justice to Peter's long-standing kind support and inspiration.

The Woolhope Naturalists' Field Club Geology Section

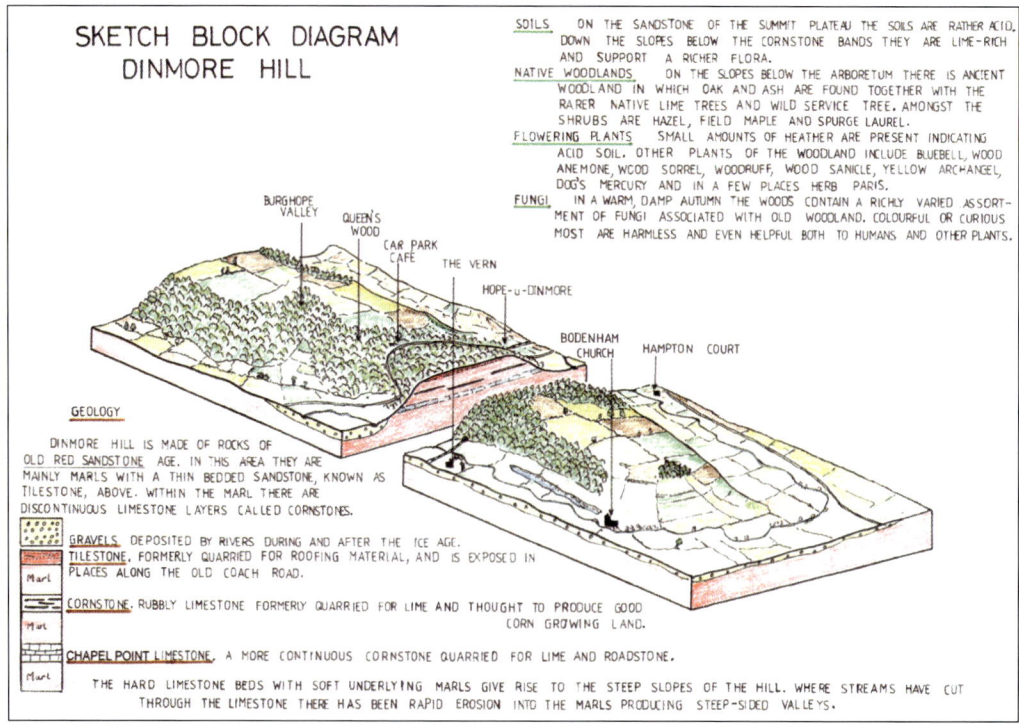

This sketch by Peter Thomson was used in his teaching of geology. Other similar sketches by him are shown in Chapter 2. All of them are updated for stratigraphical names.

Foreword

It gives me great pleasure to have the honour of writing the foreword to this book on the county's geology. My family have some claim to be the earliest promoters of geology in the area. In 1833, my forbears James Davies, Sir George Frankland Lewis and Richard Banks amongst others wrote urging Sir Roderick Impey Murchison to publish what became *The Silurian System* and promising to purchase copies 'providing that the cost did not exceed five guineas'. They were as good as their word and their names are listed among the subscribers. Sadly, all their copies disappeared from our library and I had to purchase a replacement some years back.

R.W. Banks at about the time of his collaboration with R.I. Murchison (Fig. 2.11)

However, the greatest contributor to geological work was my great-grandfather, Richard Banks's son Richard William Banks (RWB). RWB can be described as a Victorian polymath although he would only recognise the term Natural Historian. His interests included gardening, archaeology and maybe photography as well as geology; all this whilst practising as a lawyer and banker in the days when banking was regarded as respectable and left plenty of time for other occupations. It is tempting to regard him as an amateur, which in the Victorian sense he was, not being paid or formally trained in the subject, but we need to remember that there was no such thing as geological education in those days and Murchison was trained as a Royal Engineer, not a geologist! RWB's approach and that of his local associates was thoroughly scientific as an 1856 letter from Murchison demonstrates: 'I hope to see the *Pterygotus* beds with my own eyes now that you have given them a new geological interest'. RWB also sent *Pteraspis banksii* specimens to J.W. Salter at the Museum of Practical Geology in Jermyn Street.

He collected extensively in Radnorshire, Herefordshire and Shropshire with many sites in the immediate area around Kington. Many of his collections remain in the care of The Hergest Trust and are available to scholars and others interested in the subject. They have recently been catalogued and conserved with advice from Professor Derek Siveter. However, one type-status eurypterid *Pterygotus banksii* is on loan to the National Museum of Wales, Cardiff.

RWB was notorious for not joining national organisations and was never a member of the Geological Society where in 1855 Murchison himself had communicated RWB's paper on *Downton Sandstones in the Neighbourhood of Kington*. He was however a passionate early member of the Woolhope Club and was President in 1860.

Silurian geology was at the centre of the scientific world in the 1850s but gradually faded into the background until Derek and David Siveter's work on the Lagerstätte in the last 20 years brought it back to prominence. How delighted RWB would have been.

Lawrence Banks

Preface

The need for a comprehensive book on the geology of Herefordshire has been felt for some years. Requests for advice on local geological publications are frequently made. This book is the response by the Woolhope Naturalists' Field Club to satisfy this demand.

The recent growth of interest in geology amongst the general public can perhaps be attributed to the many excellent television programmes devoted to earth sciences in the last decade or so, with their spectacular images of volcanoes, landscapes and other geological phenomena. However, this follows a century of decline, at least locally, in public interest in geology.

This decline and the subsequent rebirth of the subject may be followed in the *Transactions* of the Woolhope Naturalists' Field Club. The strength of geological interest at any time can be gauged by the publication rate on this topic in the *Transactions*. The graph shows the number of such publications in each decade. After a strong beginning following the Club's formation in 1851, the rate was falling away by the 1880s but was re-invigorated in the 1890s by a number of geologically active newcomers to the Club. Thereafter, the publication rate diminished steadily, with expected low points during the World Wars, until the 1980s, when only two items appeared. The rebound in publication rate after 2000 is due almost entirely to the members of the newly formed Geology Section of the Club. The inclusion of the substantial articles of local interest in the Section's annual newsletter brings the total number in Woolhope Club publications back to the levels of the 1900s.

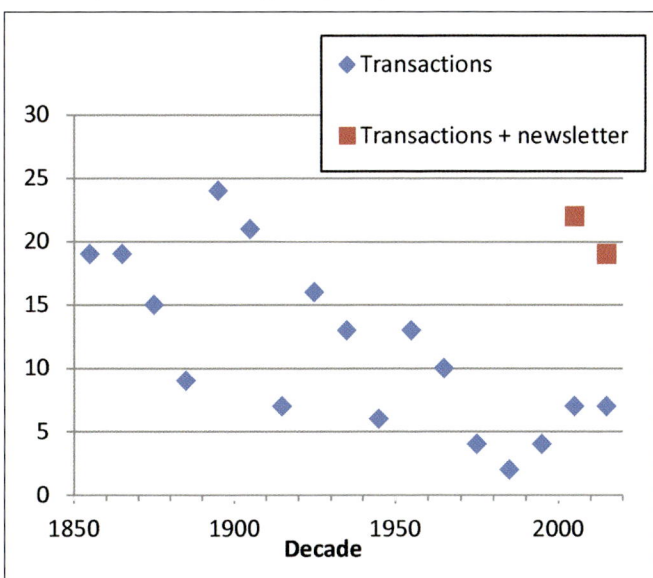

The number of local earth science items in Woolhope Club publications during each decade.

It is this resurgence of geological interest within the local community that has spurred the production of this book. No such general geology book for the county has been produced in the 200 years of modern geological science. Several short articles have been written, some in the Club *Transactions* and some as book chapters, often as introductions to books on other topics, but many of them are now out of date, being penned prior to the advent of geology's unifying theory, that of Plate Tectonics.

The south-west Midlands area appears to have been subject to a particularly strong burst of geological enthusiasm in the local community. This was stimulated initially by the formation, in 1996, of the Herefordshire and Worcestershire RIGS Group, a body run by professional geologists but with considerable support by keen non-professionals. It had the principal aim of establishing a network of Regionally Important Geological Sites (RIGS) in the two counties. This was the local part of a nationwide effort to strengthen geological conservation to parallel wildlife conservation. The name of the group was later changed to the Herefordshire and Worcestershire Earth Heritage Trust and the aims broadened to include work to enhance the public awareness of earth science.

In the following decade, amateur geological interest flourished locally. The year 2002 saw the re-establishment of a group devoted to geology within the Woolhope Naturalists' Field Club, based in Hereford. The Club, founded in 1851, initially had a major interest in geology and included some of the principal British geologists of the time as honorary members. Indeed, the name of the Club stems from the village of Woolhope which lies at the centre of an area of major geological interest near Hereford. At that time geology ranked high in public consciousness with the conflict between religion and science concerning evolution and the origin of the Earth. Several other county-based naturalists' clubs were formed locally in that period.

This establishment of the Geology Section of the Woolhope Club, the originators of this book, was followed by the creation of Geology groups of the University of the Third Age in Malvern and some Herefordshire towns, and the formation of the Teme Valley Geological Society. In addition, the Abberley and Malvern Hills Geopark was founded in the same period and extends into eastern Herefordshire. All are thriving groups with active programmes.

This book is intended as a help to those wishing to know better the origins of the Herefordshire countryside. It is increasingly recognised that the character of a locality, the vegetation, the fauna, the building stones for instance, are all greatly dependent on the rocks on which they stand. Knowledge of the geology is essential for a full understanding of a local environment.

The book is not, however, intended to be a comprehensive field guide to the geology of the county. That would be a much larger book and such a need is met, in part and for limited areas, by a range of professional publications mentioned as Further Reading at the end of each chapter and as Earth Heritage Trust and Geological Survey publications. Nevertheless, many specific locations and their significance in the landscape are indicated in the book and their National Grid References are given so that they may be readily found.

Readers already with knowledge of the geology of Herefordshire will quickly notice the use of many unfamiliar names for the various rock strata. The authority for naming these strata in the UK rests with the British Geological Survey. This organisation revises the rock names, when appropriate, in the light of recently gained knowledge. In the past ten years, many of the local names have been formally changed. It is expected that the new names will come into general use within the lifetime of this book, while the older names will gradually become disused. A table of the correlations between the old and the new is to be found in each chapter where it is relevant.

A couple of points of detail in reading the book are as follows: National Grid References (NGRs) are given with either two, three or four digits accuracy (i.e. to 1km, 100m or 10m) depending on the size of the site in question; Geological times are given with units of either Ma or My. Both represent a period of one million years. A number with 'Ma' as a unit represents a date, a fixed time in the past, the number of millions of years before the present. Those with 'My' as a unit are simply a length of time. Thus, for example, the length of time between 345Ma and 325Ma is 20My.

The dates of geological events are given in various places throughout the book. These are subject to revision in the light of later research, as decided by the International Commission on Stratigraphy. (See http://www.stratigraphy.org/index.php/ics-chart-timescale for the current data.)

It is necessary to state here a number of warnings concerning site visits: many of the sites in the book are on private land and permission must be sought to visit them; it will usually be given but cannot be assumed. Nor does mention of a site imply that it is safe to visit. Almost all geological sites have some degree of danger, even if slight. Visitors are always responsible for their own safety and that of others.

Sample collection must be governed by the widely published code of conduct for geologists. Collecting from within SSSIs is not allowed. In Herefordshire, that particularly means the Malvern Hills. Some other sites are of very limited extent or contain easily damaged features. These must not be disturbed. Take photographs rather than rocks or fossils.

*Herefordshire – A land of rivers and Old Red Sandstone
seen here in the River Wye and Brobury Scar
(© John Payne)*

Contributors

Kate Andrew
Kate has spent most of her career in museums, maintaining curatorial responsibility for geology collections in addition to progressively more senior management roles. From her first forays into museums, she has been fascinated by the history of natural science in the late 18th to mid-19th century and the explosion of interest in collecting. It was this interest, initially in geology, that led directly to the establishment of provincial museums by local scientific societies.

Kate was County Curator of Natural History based at Ludlow Museum from 1995 to 2002 and Principal Heritage Officer for Herefordshire from 2002 to 2012. She has also undertaken conservation condition surveys of all major West Midlands geology collections and undertook the computer cataloguing of the geology collection at Dudley Museum. She is therefore very familiar with the historic geology collections of the region and the geology of Herefordshire and Shropshire in particular.

Gerry Calderbank
Gerry worked for 35 years in educational and academic publishing, originally with Methuen Educational and subsequently with Cambridge University Press. Early retirement provided more opportunity for leisure interests, especially canoeing, rambling, geology and archaeology. He founded the Wye Valley Branch of the Canadian Canoe Association and has held most offices of the Hereford County Canoe Club.

Gerry joined the Woolhope Club in 1963 and was a co-founder and then third Chairman of the Woolhope Archaeological Research Section. He is a Trustee of the Herefordshire Archaeological Trust, and was likewise with the Earth Heritage Trust, which he briefly chaired during its transition to Charitable Limited Company status.

As co-instigator of the Woolhope Geology Section, he was its founding Secretary and then Chairman in turn. Gerry has an interest in the geology and archaeology of transport systems, especially the Leominster Canal, and is Secretary to the Friends of the Leominster Canal.

Derek Foxton
Derek attended local infant and primary schools, then Ledbury Grammar School 1952-1959. After graduating from the University of London, Royal Dental Hospital, in 1964, he joined a practice in Commercial Street, Hereford. In 1967 he purchased it and then managed a multi-surgery practice for 36 years.

His hobbies include collecting historic motorcycles and associated literature. He is a worldwide consultant on several early manufacturers.

His interest in photography started while at school. Saturday mornings and holidays were spent in the 'Derek Evans Photography' studio, Broad Street, Hereford. This was a freelance national news and photo-journalism business. This led to five years of semi-professional work for the *Daily Express*, while at university. Over the past 50 years he has photographed many nooks and corners of Hereford. During this time he collected and copied thousands of old photographs and postcards of the city and county. Using his photographic archive and library, Derek has written nine books on the area. After photographing highlights of the county during various flights, he published *Herefordshire from the Air*, a book in colour, in 2000. His recent book *Hereford through Time* was published by Amberley Press.

Dave Green

Dave has spent most of his career teaching Geology. He discovered the subject whilst teaching Geography and has been hooked ever since, with an insatiable desire to know more about all aspects of our fascinating discipline. In particular he loves field geology, the understanding of the relationship between the underlying rocks and the scenery above, and the challenge of making sense of the whole from the sum of a very small percentage of the parts.

Dave spent many years as Head of Geography and Geology at Stroud College, from 1996 as Head of Geology at Sir Thomas Rich's in Gloucester, from 1998 to 2010 teaching extra-mural classes for Bristol University, from 2001 teaching A-level and recreational classes at John Kyrle School in Ross, and runs his own adult classes and trips under Geostudies.co.uk. Although living (just!) in Gloucestershire, he has spent a great deal of geological time in Herefordshire.

Charles Hopkinson

Charles is a retired farmer and latecomer to geology. Having searched in vain for an accessible, up-to-date guide to Herefordshire's geology, he suggested that members of the Woolhope Club should write this book. As an amateur historian with a particular interest in the brittle relationship between medieval England and Wales, he is intrigued by the influence of geology on the ebb and flow of the protracted Anglo-Norman conquest of Wales. The geology of Wales with its mountainous terrain had by the 11th century contributed to the division of the country into a number of small, independent or semi-independent, geographically isolated states, occasionally unified for a time under strong central leadership. These small states were in many cases often relatively easy prey for Anglo-Norman conquistadors who could pick them off one by one and establish their lordships – an example of a connection, however remote, between geology and history.

He is currently a committee member of the Woolhope Geology Section committee.

Moira Jenkins

Moira graduated in geology at Aberystwyth University in 1966. She worked as a geophysicist in an oil exploration company and then returned to Aberystwyth University to work as a map curator and research assistant, investigating the sub-Pleistocene topography of the

Irish Sea. After completing a Postgraduate Certificate of Education, she taught primary school children and then married and brought up her family, doing various jobs such as helping at play group, further teaching and working as a court usher. Since 2002 she has worked at Herefordshire and Worcestershire Earth Heritage Trust, writing geological trail guides, producing interpretation panels, recording geological sites, producing the Herefordshire Geodiversity Action Plan, leading guided walks and increasing public awareness of Herefordshire's wonderful and interesting geology. Moira is the geology recorder for the Woolhope Naturalists' Field Club and would like to hear of any temporarily or newly exposed geodiversity sites.

Dr Paul Olver FGS FRAS
Since completing his doctorate in geology at the University of Birmingham where he studied the volcanoes of southern Italy, Paul has spent his whole career in education and, since 1974, within local authority adult education services in both Surrey and Herefordshire. He now works part-time as an educational project consultant and has continued to tutor both day and evening courses in geology and astronomy. He now has more time to indulge in his leisure interests including his passion for railways both life-size and for modelling. He is also currently Chairman of the Herefordshire Astronomical Society and the dark skies of the county have encouraged him to set up his own observatory.

Paul joined the Woolhope Club in 2000 and was the co-founder and first Chairman of its Geology Section. He is also a Trustee of the Herefordshire & Worcestershire Earth Heritage Trust whose geoconservation and community education work all fit in well with his overall aim of promoting earth sciences to the general public.

Dr John Payne
John grew up in Weymouth, took a degree in physics at Imperial College and worked in Malvern as a physicist for the Ministry of Defence. On retirement he took up his long-standing interest in geology and joined what is now called the Herefordshire and Worcestershire Earth Heritage Trust (EHT), based in the University of Worcester. He was for a period the Chairman of EHT and is now a Vice-President. He was a founder member of the Geology section of the Woolhope Naturalists' Field Club and is currently a committee member. He is a member of three other geological organisations. For all of these he leads geological walks locally and gives occasional talks on aspects of local geology. He is currently active in organising a team of volunteers to clean up important local geological sites and is the editor of this book.

John Stocks
John spent his career as a Mining Engineer. His early years were in the Nottinghamshire coalfield at a time when conventional picks and shovels were the norm. Later he became closely involved in the development and application of mechanized mining. Working for several mining companies, this resulted in considerable time spent overseas and during his career he worked in more than forty countries in a wide range of geological, political and climatic settings.

On retirement John studied Earth Sciences with the Open University, graduating in 2010. It was natural therefore that his interests led him to work with the Herefordshire and Worcestershire Earth Heritage Trust (EHT). Here his interest in photography and landscape was rewarded by working on the Abberley and Malvern Hills Geopark Way and Champions projects. John was later encouraged to present a workshop on geophotography: 'Rocks through the lens'. This Woolhope Club book continues the journey.

Robert Williams
As a geographer, historian and long-standing member of the Woolhope Club Geology Section, Robert Williams has particular interests in the influence of geology and geomorphology on the natural landscape and the history of geology as a science – especially with reference to the 'Marches' area, the border country of England and Wales which includes Herefordshire. His degree in History and Geography from Cambridge led him into education. He has taught both these subjects at secondary and sixth-form level and, in recent years, has been involved with adult education. He has jointly run Summer Schools in north-west Herefordshire on a range of themes relating to geology and landscape in the Welsh Marches and has led or contributed to the work of two local U3A Geology Groups. He has also worked as one of a small team with the Geology Champions scheme set up by the Herefordshire and Worcestershire Earth Heritage Trust (EHT).

Publications from The Earth Heritage Trust and The British Geological Survey

All of the authors of the chapters in this book recommend some additional information sources which the reader may find useful. Amongst these, the publications of the Herefordshire & Worcestershire Earth Heritage Trust and the British Geological Survey are always included so they are listed here to avoid considerable repetition later on. Where especially relevant, some are specifically referenced within certain chapters.

Herefordshire & Worcestershire Earth Heritage Trust (H&W EHT) has published a series of 25 'Explore' trail guides. Those for Herefordshire are listed below, with the relevant book chapters.

Landscape and geology trail guides	
Symonds Yat	Chapters 8 and 9
Woolhope Dome	Chapters 2, 6 and 8
Wye Gorge	Chapters 2, 8 and 9
Malvern Hills	Chapters 2, 4 and 6
Byton & Kinsham	Chapter 9
Hampton Bishop	Chapter 9
Wyche & Purlieu	Chapters 4 and 6
Queenswood & Bodenham	Chapters 7 and 9
Ross-on-Wye	Chapters 7 and 9
Kington & Hergest	Chapters 2, 6 and 9
Wigmore Glacial Lake (out of print)	Chapter 2 and 9
Building stones trail guides	
Goodrich Castle	Chapters 7 and 10
Hereford Cathedral	Chapters 1 and 7
Hereford City Centre	

Also very useful is the larger 'Frome valley discovery guide' Chapters 2, 6 and 9

These EHT guides are written in generally non-technical language. They are often available for sale at Tourist Information Centres as well as directly from the Earth Heritage Trust (Geological Records Centre, University of Worcester, Henwick Grove, Worcester WR2 6AJ; eht@worc.ac.uk; 01905 855184; http://www.earthheritagetrust.org).

Authoritative professional accounts of local geology are available in the output of the British Geological Survey (BGS). The BGS produces a series of geological maps for the county at 1:50000 scale. (The coverage of the maps is complete except for a part of northwest Herefordshire. For this area, the only full map dates from the 1800s. Non-BGS work is at present under way to remedy this lack.) The maps also contain geological cross-sections. They may, of course, be purchased, but can be viewed on-line at http://www.bgs.ac.uk/data/maps/home.html. Each map has an associated technical document, either as a detailed memoir or as a simpler explanation of the geology. These BGS publications assume a certain familiarity with geological concepts and technical terms. The relevant map sheets and descriptive publications are listed below.

Ludlow	Sheet no. 181 – Map only
Hay-on-Wye	Sheet no. 197 – Map and sheet explanation
Hereford	Sheet no. 198 – Map; Memoir for Hereford & Leominster
Worcester	Sheet no. 199 – Map and memoir
Talgarth	Sheet no. 214 – Map and sheet explanation
Ross-on-Wye	Sheet no. 215 – Map; Sheet explanation for Hereford & Ross-on-Wye
Tewkesbury	Sheet no. 216 – Map and memoir
Monmouth	Sheet no. 233 – Map; Memoir for Monmouth & Chepstow

Technical descriptions of many specific geological SSSIs may be found in the volumes of the Geological Conservation Review, published by the Geologists' Association since 2011 and previously by JNCC. Volumes containing details of Herefordshire sites are listed below. Several of these are available on-line (http://jncc.defra.gov.uk/page-2731).

Volume No. 9 (1995)	*Palaeozoic Palaeobotany of Great Britain*, Cleal, C.J. & Thomas, B.A.
Volume No. 13 (1997)	*Fluvial Geomorphology of Great Britain*, Gregory, K.J. (Ed.)
Volume No. 16 (1999)	*Fossil Fishes of Great Britain,* Dineley, D. & Metcalf, S.
Volume No. 18 (2000)	*British Cambrian to Ordovician Stratigraphy*, Rushton, A.W.A. *et al.*
Volume No. 19 (2000)	*British Silurian Stratigraphy*, Aldridge, R.J. *et al.*
Volume No. 20 (2000)	*Precambrian Rocks of England and Wales*, Carney, J.N. *et al.*
Volume No. 31 (2005)	*The Old Red Sandstone of Great Britain*, Barclay, W.J. *et al.*
Volume No. 35 (2010)	*Fossil Arthropods of Great Britain*, Jarzembowski, E.A. *et al.*

1 AN INTRODUCTION TO HEREFORDSHIRE

One of the most remarkable facts about the early years of the 19th century is the emergence of an accurate geological map of the British Isles. Today, when we go 'into the field' to study geology, we take a geological map with us to help our studies, but this was not possible for the early pioneers of the new science of geology, since such maps did not exist. Yet, when it came to producing them, these practical and scholarly people not only had to map out the 'solid geology' (i.e. the underlying rock) but to do so by looking out over a landscape which was principally the 'green and pleasant land' that William Blake (1757-1827) had spoken of – one covered in trees and other vegetation. In general most of lowland Britain does not have bare landscape with large areas of exposed rock, and this is certainly true of Herefordshire, although the rocks are more exposed in the upland areas of the north and west of the country. Two hundred years ago, however, William Smith (1769-1839) produced a map that was essentially correct for England, if rather vague for the older and as yet unclassified rocks in much of Wales and Scotland. By 1839, in his masterly work *The Silurian System,* Roderick Murchison (1792-1871) had managed to work out the detail of the older geology of Wales and the Welsh Marches. Much of Murchison's pioneering work was done in Herefordshire and still forms the basis for modern descriptions of the county's geology. The maps that accompanied his great work are remarkably like the present day offerings from the British Geological Survey (BGS) (Fig. 1.1). By virtue of their 'eye for the

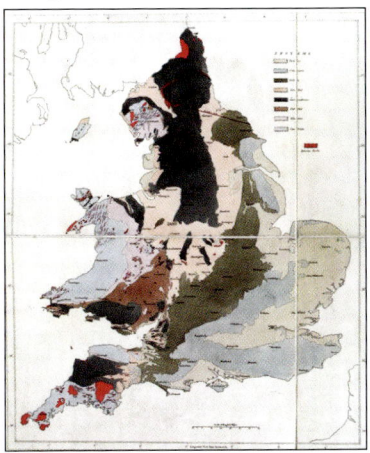

*Figure 1.1   Illustrations of (left to right) Smith's map (1815),
Murchison's map from* The Silurian System *(1839) and BGS digital map (2015). (BGS map reproduced with the permission of the British Geological Survey ©NERC. All rights reserved)*

land' early geologists gradually developed a scientific understanding of the rocks beneath their feet. Such insights into the landscape remain valuable today. It is hoped that this book will assist readers in their own understanding of the geology and landscape.

The many later researchers have, of course, added very greatly to Murchison's original picture. This book aims to give an account of the up-to-date knowledge of the geology of one of Britain's most distinctive counties: Herefordshire – a county with a dramatic geological story to tell and one which has been traversed both by distinguished geological pioneers in the early 19th century and countless scholars and interested amateurs ever since.

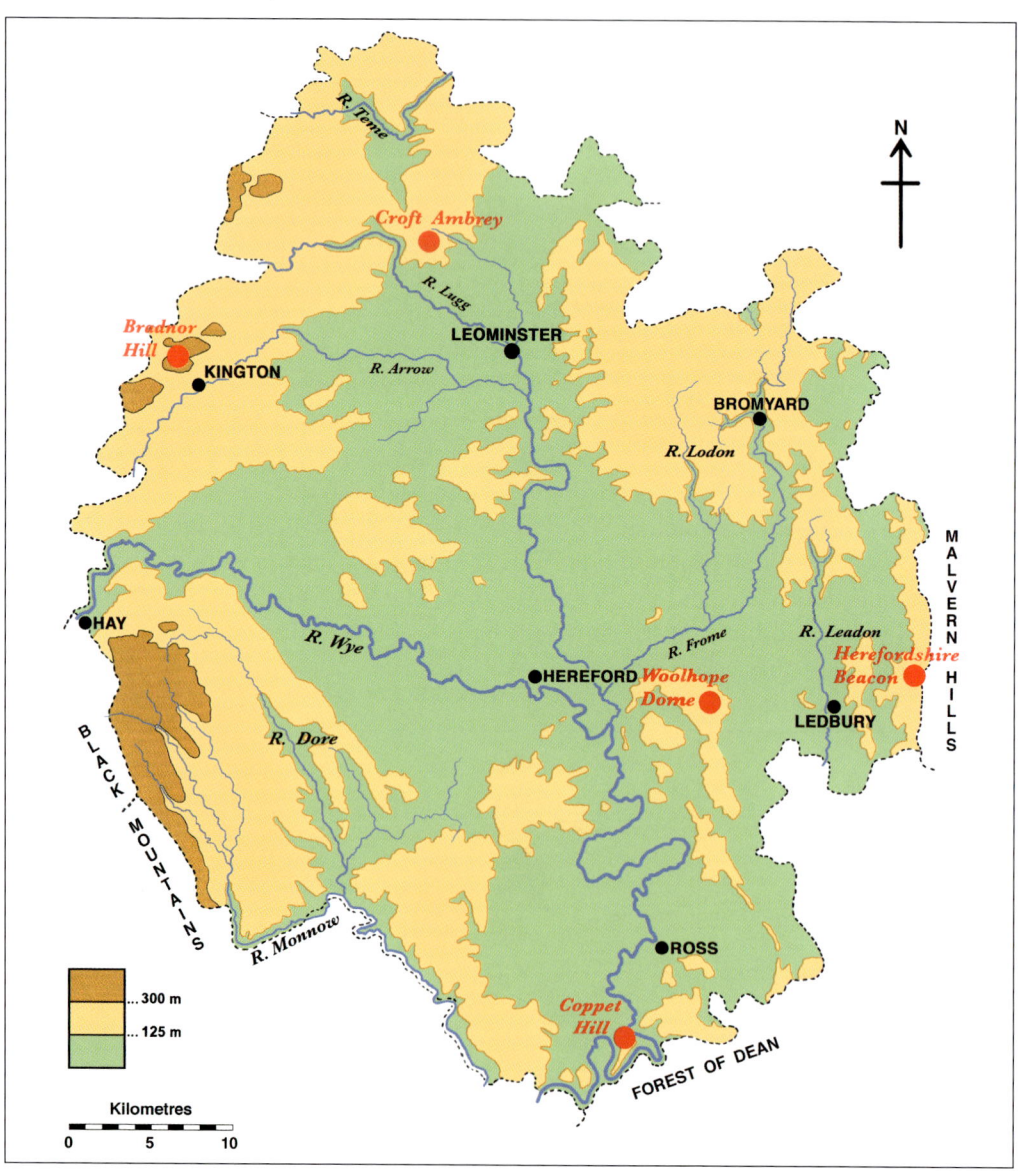

Figure 1.2 Herefordshire: relief map showing the 'Five Hills' of the following chapter, Hereford City and the main county towns. (Drawn by Gerry Calderbank)

Landscape and topography of the county

All roads lead to Hereford – a commonplace no doubt for all county towns as well as Rome! But a brief glance at a map of the county shows Hereford set conspicuously at the centre with a series of roads approaching from all points of the compass. Equally striking is the distribution of market towns around the capital city – Kington, Leominster, Bromyard, Ledbury, Ross-on-Wye and Hay-on-Wye – each of them between 14 and 20 miles from Hereford. This settlement pattern says much about the topography and associated geology of Herefordshire (Fig. 1.2).

The county consists substantially of a 'basin' or central plain, with upland areas on the margins. A topography such as this, with a uniform plain, allows for the free movement of transport across it and thereby the theoretical likelihood of the development of a hierarchy of 'central places' as the German geographer, Walter Christaller, pointed out.[1] In practice, various hills rise from the plain but none is so dominant as to impede the 'free movement' mentioned above. Rising above the Herefordshire plain are the 'cornstone' hills, of Devonian rock, remnants left by the erosion of surrounding beds. These are particularly evident at Dinmore Hill between Leominster and Hereford (see diagram on p.*viii*); Burton Hill in the north of the county between Mansel Lacy, Weobley and Wormsley; Credenhill, an Iron-age hill fort to the north-west of Hereford City; and the conically shaped Pyons, which give their name to the villages of King's Pyon and Canon Pyon.

Also significant in land which rises above the plain is the area between Hereford and Ledbury in the eastern part of the county, known to geologists as the Woolhope Dome. Centred on the village of Woolhope, this up-folded structure has been 'revealed' by the erosion of younger rocks above it (see View no. 5 in Chapter 2).

Across this landscape flow two main river systems, with a small one in the east. In the north of the county, the River Teme flows east along the Herefordshire-Shropshire border, passing round Ludlow and eventually joining the River Severn at Worcester. For the major part of the county, the River Wye provides the main system of drainage – a system which can be broadly divided into three main sections:

> In the north, the Lugg with its tributaries the Arrow and the Frome. The Lugg joins the Wye to the east of Hereford at Mordiford, on the western edge of the Woolhope Dome.
>
> The central artery of the Wye itself, which flows into the county from the west, via Hay-on-Wye. Hereford city lies on the Wye, just before the confluence with the Lugg. After this, the river's course goes south to Ross-on-Wye and the Wye Gorge before it passes out of the county on its southerly passage to the Bristol Channel at Chepstow.
>
> In the south, the Monnow and associated tributary rivers such as the Dore flow south-east from the Black Mountains to join the Wye at Monmouth. The Monnow forms the south-western boundary of the county.

In addition, to the east is the small south-flowing River Leadon which passes through Ledbury on its way to Gloucester.

Geology of the county

Within the context of the British Isles as a whole, Herefordshire displays a sequence of rocks ranging in age from 677 to 245 million years (from the late Precambrian and the Palaeozoic Eras; see Chapter 3). By comparison with the age of the earth (4,600 million years) they are quite young but they are distinctly older than almost all of the rocks of central England, which fall into the Mesozoic Era (245 to 65 million years ago).

In relatively recent times, the county has been affected by successive glaciations, which took place in the Quaternary Period, beginning about two million years ago.

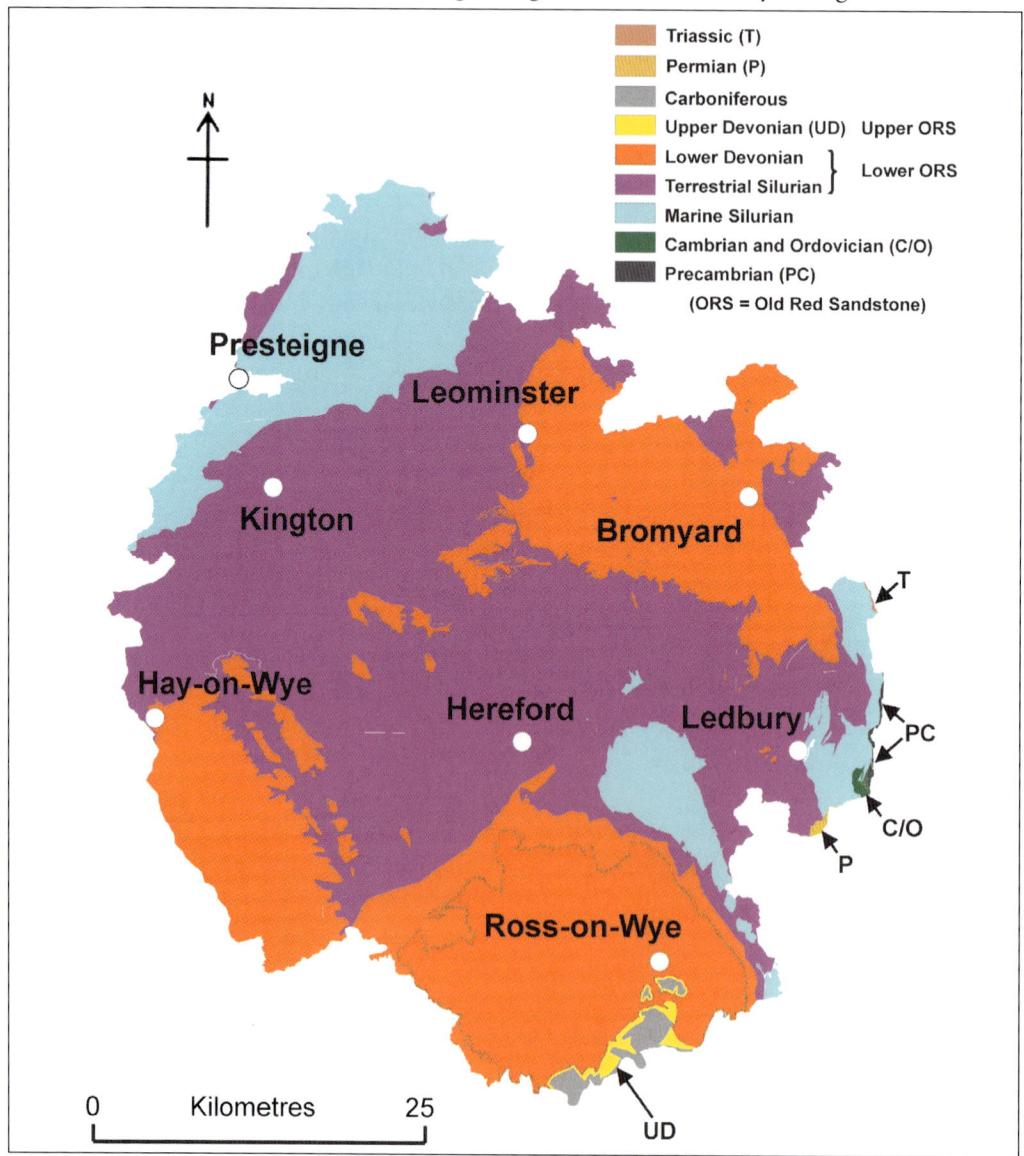

Figure 1.3 Geological map of Herefordshire. The key shows how the rocks are divided into differently aged groups, with the oldest at the bottom in the key. The actual age range of each group, or geological period, is shown in Fig.3.1. (Drawn by Gerry Calderbank)

The map of the solid geology of Herefordshire shown in Figure 1.3 illustrates the distribution of the rocks of these various systems across the county.

The detailed description of Herefordshire's geology is, of course, the main subject of this book and is given in Chapters 4 to 9 but two further overall views of the county's geology appear later in the book (Figure 6.5 with its associated text and in the Introduction in Chapter 9) and may be read in conjunction with Figure 1.3.

Much more detailed information, including both solid and superficial deposits such as glacial material, is to be found on the British Geological Survey map sheets listed on the page 'EHT and BGS Publications' (pages *xix-xx*).

The Old Red Sandstone in Herefordshire

The Old Red Sandstone has been called 'the Herefordshire stone.'[2] Rocks of the Old Red Sandstone (ORS) underlie most of the area of Herefordshire (Fig. 1.3) and give a distinctive red colouration to many of the soils and building stones. A particular example is the fine Norman cathedral of St Mary and St Ethelbert at the junction of Broad Street and King Street in the centre of Hereford city (Fig. 1.4). Looking resplendent in a setting sun, it shows off to advantage not only the skill of its medieval craftsmen, but also, in the oldest parts, the Old Red Sandstone, hewn from quarries at Capler (near How Caple) to the south of the city and transported to Hereford by cart and barge.

Across the road, on Broad Street itself, is the entrance to the Museum, Public Library and the Library of the Woolhope Naturalists' Field Club (Fig. 1.5). In this building in the

Figure 1.4 Hereford Cathedral, built mainly of Old Red Sandstone. (© John Stocks)

19th century significant contributions were made to the early development of geology as a science by the local work of the members of the Woolhope Club. Their work in geology continues today, exemplified by the production of this book.

The high proportion of Old Red Sandstone in Herefordshire is shared by few, if any, other British counties but large areas are found elsewhere in south-east Wales and the Marches, and in northern and central Scotland. The wide separation of the two main groups is a consequence of the geography at the time of their formation, described in Chapter 7. Initially, a mountain range was centred on southern Scotland and the erosion products were washed to the north and the south, forming the two main areas of the ORS. Quite soon after, mountains arose in north Wales and the erosion of these provided most of the material for the Herefordshire ORS. The climate, arid and sub-tropical, was much the same in both. The major rocks of Herefordshire thus bear similarity to those of Orkney and Caithness. This shows in the local building stones; St Magnus Cathedral in Kirkwall has similarities to Hereford Cathedral. Fossils of many of the same creatures are found in both areas but are more rare in the south. This was a source of disappointment to the early Woolhope Club geologists who viewed with some envy the fine fossil fishes discovered by their Scottish contemporaries.

Figure 1.5 Entrance to Hereford Museum, Public Library and the Woolhope Club. (© John Stocks)

The landscape of Herefordshire, its hills, valleys and rivers, is the result of a series of geological processes. Some of these were active hundreds of millions of years ago in forming the present rocks from pre-existing materials. Later, other processes emplaced the rocks in their current positions. In relatively recent times, within the last million years and continuing to the present, erosion processes have produced the details of the landscape of today. Each of these actions will continue, in some form, into the distant future, with the 'short'-term results of a continual lowering of the hills by erosion and the effects of a probable new ice age in a few thousand years' time – assuming we are currently in a warm inter-glacial period. In the very long term, fresh episodes of mountain building or submersion in the sea will render the area unrecognisably different from today.

In the meantime, we may take the opportunity to view the county in its present picturesque form. An excellent way to achieve this is to study the scenery from the peaks of some of the high hills on the edges of the county. Five such viewpoints are identified and their stories are told in the following Chapter.[3]

2 Views from Five Hills

Before the more detailed account of the geology given later in the book, in this chapter we will describe the first impressions conveyed by the contemporary landscape from some distinctive viewpoints. Herefordshire is blessed with a magnificent landscape and the visitor – whether arriving from north, south, east or west – can easily find upland areas from which to view the wider prospect. This chapter explores the geological and landscape significance of views from five main hills in the county, four on the periphery and one more central, establishing their geological setting and the views from each of them. These hills are identified in Fig. 1.2. Geological trail guides for some of the locations are available from the Herefordshire & Worcestershire Earth Heritage Trust (EHT). (Listed on pages *xix-xx*, 'EHT and BGS Publications'.)

View No. 1: *The Herefordshire Beacon*: **approaching Herefordshire from the East**
Three of the main roads into Herefordshire from the east (the A4103, the A449 and the A438) pass close to the Malvern Hills, the A449 passing through Great Malvern and Ledbury. Crossing the county boundary on this road, a good vantage point from which to view the county is the Herefordshire Beacon, on which is the Iron Age hillfort known as British Camp. There is a car park by the main road and a well-marked footpath from there to the summit (SO 760 400).

The Herefordshire Beacon (height 338m above sea level) stands about midway along the north-south ridge of the Malvern Hills, slightly to the west of the main axis. To the north is the Worcestershire Beacon (height 425m above sea level). The county boundary (between Worcestershire and Herefordshire) runs north-south along the Malvern ridge.

From the summit of the Herefordshire Beacon, look north along the line of the hills, which rise majestically from the lowland to east and west. The hills themselves show the oldest exposed rocks in England. The rocks were formed deep underground in Precambrian times (700 million years ago) at a location between 60º and 70º south of the equator (see the Geological Timescale, Fig. 3.1) and were raised to the surface about 300 million years ago. As such they stand proud and ancient, much more ancient than the 'ancient Britons' of the Iron Age hillfort on which you are standing (built about 2,500 years ago and still occupied when the Romans came about 2,000 years ago). (The evolution of the Malvern Hills is described in more detail in Chapters 4 and 8).

Looking westward into Herefordshire, the view is of a rolling countryside, with a series of ridges marking mainly east-facing escarpments of hills which dip westward towards the centre of the county.

Figure 2.1 English Air - South Prospect by David Prentice (1998): reed pen and watercolour. View of the Malvern Hills looking south. The Herefordshire Beacon is on the right (west) in the distance. (© David Prentice)

Figure 2.2 View of the Malvern Hills looking north from the Herefordshire Beacon, showing that this hill is out of line with the main ridge. (© John Stocks)

In the valley below the Beacon is a wonderful example of geology revealed by a landscape feature. A narrow band of woodland is seen curving out across the valley. The wood stands on a ridge, the Ridgeway, which is the exposed edge of a limestone layer bent upwards (an anticline), tilted to the north and then eroded along with the overlying softer rocks to show this curved upstanding ridge of the most resistant of the rocks. The Ridgeway is shown in Fig. 2.3 and similar features elsewhere in the county are shown in Figs. 2.15 and 2.26. Since the anticline is tilted down to the north, the older, lower-lying rocks come to the surface to the south. These are sandstones of the hill on which stands the Obelisk (SO 752 378), a memorial to various members of the Somers family who lived at Eastnor and owned much of the surrounding land. Out of sight beyond Obelisk Hill is an area of older rocks of 500My age. Exposures of rocks of this age are uncommon in England.

The local hills west of the Malverns are composed of limestones, siltstones and shales of Silurian age. They were laid down under tropical seas on an Ordovician and Precambrian basement when this area was about 30° south of the equator.[1] Ledbury, the first of the satellite market towns around Hereford, is set just beyond these hills on Silurian rocks of the lower Old Red Sandstone (ORS), overlooking the River Leadon which flows south to join the Severn near Gloucester. The poet John Masefield (1878-1966) was born here and reflected on the landscape and the red soils of his native Herefordshire throughout his life. When, on being appointed Poet Laureate in 1930, he was given the Freedom of the City of Hereford, he said:

> .... I am linked to this County by subtle ties, deeper than I can explain: they are ties of beauty. Whenever I think of Paradise, I think of parts of this County. ... I know of no land more full of bounty and beauty than this red land, so good for corn and hops and roses. (The full text of this speech is given in the Appendix.)[2]

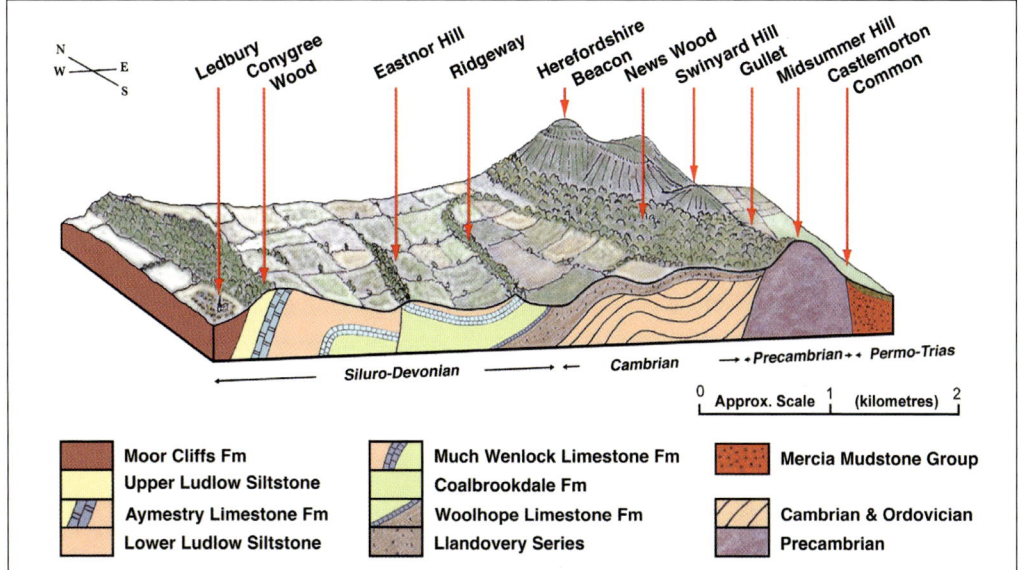

Figure 2.3 Block diagram of Ledbury to the South Malvern Hills. (Drawn by Peter Thomson with further work by Gerry Calderbank)

Beyond Ledbury, the Herefordshire plain with its familiar red soils and rocks stretches to Hereford and on westwards to the Black Mountains, which mark the western boundary of the county. These are formed of mudstones and sandstones of Silurian and Devonian age, laid down as sedimentary rocks – the result of the erosion of massive mountains to the north and west when this area was near the equator. They are often referred to as the Old Red Sandstone (ORS) to distinguish them from the New Red Sandstone of the more recent Permian-Triassic age, which forms much of the Midland Plain to the east and north of the Malvern Hills. (For the derivation of the names Silurian and Devonian see pages 91 and 123.)

This westerly view across the Silurian/Devonian Herefordshire Plain is interrupted to the west of Ledbury by an upland area. This is the famous Woolhope Dome, hills of Silurian rocks which protrude as an anticline (an upward fold of the rocks; see Chapter 3) through the younger ORS rocks of the plain. It was this formation which so fascinated the early members of the Woolhope Naturalists' Field Club in the 19th century (see View No. 5). The eastern edge of the Dome can be seen from the Herefordshire Beacon and is known as the Marcle Ridge, taking its name from the village of Much Marcle at the southern end – the home of one of Herefordshire's oldest and most famous cider firms.

To the south, the Malvern ridge continues to Chase End Hill. Beyond this, in Gloucestershire, lies May Hill, recognisable by the prominent clump of trees on its summit. This, like the Ledbury Hills and the Woolhope Dome, is another anticline in Silurian rocks. On a clear day a part of the Severn estuary can be seen in the distance. To the west of May Hill the distant ground is the northern edge of the Forest of Dean, where lies the southern border of Herefordshire.

Eastward from the Herefordshire Beacon, the view is across the flat-lying Triassic and Jurassic rocks of the Severn Valley to the Cotswold Hills. Bredon Hill, an outlier of the Cotswolds, is prominent. To its north is the Vale of Evesham and further north again are the Lickey Hills with Birmingham out of sight beyond them. This spectacular view is into

Figure 2.4 View south from the Worcestershire Beacon from an engraving by Henry Cross c.1875.

Worcestershire and so is not an immediate part of our story. The origin of this and the Malvern landscape is nevertheless highly relevant to Herefordshire's story and is described, with the Malverns, in Chapter 4.

Before leaving the Herefordshire Beacon, look back northward towards the Worcestershire Beacon. In the 1850s, the great geologist, Sir Roderick Murchison led a field excursion of the Woolhope Club to this hill (for more on Murchison see Box 2.2). There, on the highest point of the Malverns, he pronounced on the latest findings in the new science of geology. Looking east over the Severn Valley and to the Cotswolds, he could speak of the younger Mesozoic rocks (Triassic and Jurassic) and, to the west across Herefordshire, of the older Palaeozoic rocks, to which he had devoted so much of his life – particularly the Silurian and Devonian.

Today, geologists recognise an important geological fault on the eastern side of the Malverns, while to the west the land lies higher. The road westward from the Herefordshire Beacon into Herefordshire is thus of a gentler incline than the steep rise from the east.

View No. 2: *Bradnor Hill*: **approaching Herefordshire from the West**
Three main roads enter Herefordshire from the west (the A44, the A438 and the A465). The A438 and the A465 offer spectacular views of the Black Mountains, on the western edge of the county. The A44 enters into the county via Kington with access to Bradnor Hill just off the by-pass on the northern side of the town (SO 302 569). Approaching from the west, take the first left turning to Titley off the first roundabout on the by-pass. Within about 50 yards, take the first left turning and follow the narrow lane up the hill. This leads to Kington Golf Course at the top of Bradnor Hill. Continue on the road past the Club House at the top and across the Golf Course for about half a mile to the point where a notice says 'No cars beyond this point'. Park here. (N.B. Take great care on the narrow road to the top of Bradnor Hill. There are passing places but it is a single track road with farm traffic and traffic going to and from the Golf Club.) [OS Landranger 148 'Presteigne & Hay-on-Wye' and 161 'The Black Mountains']

From the car park at the top of Bradnor Hill, there are magnificent views on a clear day in all directions. To the south-east, the Herefordshire Plain is laid out showing its Silurian/

Figure 2.5 View from Bradnor Hill looking south-east over the north Herefordshire plain.
(© Robert Williams)

Devonian red soils with hills rising above the plain. Most prominent is the plateau-like wooded Burton Hill in the middle distance, and, to its left, a conical hill (the 'pyon') at Canon Pyon. These are examples of weathered Old Red Sandstone remnants, sometimes referred to as 'cornstone hills' because layers of nodular limestone within the beds of sandstone were thought to contribute to a soil type which was good for corn growing. These hills, which stand up across the Herefordshire Plain also include the Bromyard Downs east of Leominster, Dinmore Hill between Leominster and Hereford on the A49 and Credenhill, the site of an Iron Age hillfort north-west of Hereford. Hereford itself is beyond Burton Hill.

Bradnor Hill is part of a ridge of upper Silurian rocks, mainly mudstones and calcareous siltstones. The ridge extends across the north-west of the county, emerging from under the younger rocks of the Herefordshire Plain. Looking north-eastwards along the ridge, it is seen to include Wapley Hill, Shobdon Hill, Croft Ambrey (see View No. 3) and Bircher Common, all with rocks dipping to their right (south-east). The Mortimer Trail walk follows this line of hills.

The same Silurian ridge extends south-westward across the golf course to Hergest Ridge above Kington. On its south-eastern edge are the distinctive gardens of Hergest Croft, belonging to the Banks family. The gardens are open to the public and in May and June have spectacular displays of azaleas and rhododendrons (Fig. 2.6), lovers of lime-free soils – an interesting reflection of the underlying geology, which at this point displays a transition between the calcareous marine rocks of Hergest Ridge and the terrestrial Old Red Sandstone of the Herefordshire Plain. (This area is partly covered in EHT Trail Guide *Kington & Hergest*.)

Looking south-east again, on a clear day the Malvern Hills can be seen in the distance, with the Herefordshire Beacon towards the southern end. The Silurian outcrops on the

---

Box 2.1   Local Landowners and the Picturesque Artistic Movement

During the 18th century, the owner of the Foxley estate on the western side of Burton Hill, Uvedale Price (1747-1829), looked out towards Bradnor Hill, the Black Mountains and Wales beyond and, reflecting on the magnificence of the landscape across the north Herefordshire plain, initiated ideas about the 'Picturesque' – a theme also being developed by Richard Payne Knight of Downton (1751-1824), at one time Price's neighbour at Wormsley on the eastern side of Burton Hill. This idea was also developed by the Revd William Gilpin (1724-1804) in the lower part of the River Wye, south of Ross-on-Wye (see Boxes 2.5 and 10.1). In due course, ideas of the Picturesque would feed into the Romantic Movement, but in its origins, this aesthetic was closely associated with Herefordshire and its Palaeozoic geology.[3]

It was Downton Gorge and its landscape, which inspired Richard Payne Knight, friend and associate of Uvedale Price at Foxley, to his thoughts on 'The Picturesque' artistic movement (see Box 10.1). A generation earlier it had been on this stretch of the Teme that Payne Knight's uncle, Richard Knight, had established one of the early iron-works of the Industrial Revolution, drawing iron from Clee Hill, charcoal from the local forests, limestone from the local Silurian rocks and water-power for the forges from the River Teme as it poured through the gorge.

Figure 2.6 Rhododendrons at Hergest Croft Gardens. (© John Stocks)

north-west of the county are the counterparts to similar outcrops west of the Malvern Hills near Ledbury. To the south-west lies the prominent northern scarp edge of the Black Mountains (Fig. 2.7). This is probably Herefordshire's most prominent landscape feature. Within the upland area of the Black Mountains, most of which is outside of Herefordshire, a series of valleys has been cut by rivers flowing south-eastward as part of the Monnow river system (Fig. 9.10). Only the nearest valley, a small one, the Olchon valley, is within

Figure 2.7 The Black Mountains: the south-westerly view from Bradnor Hill. (© Robert Williams)

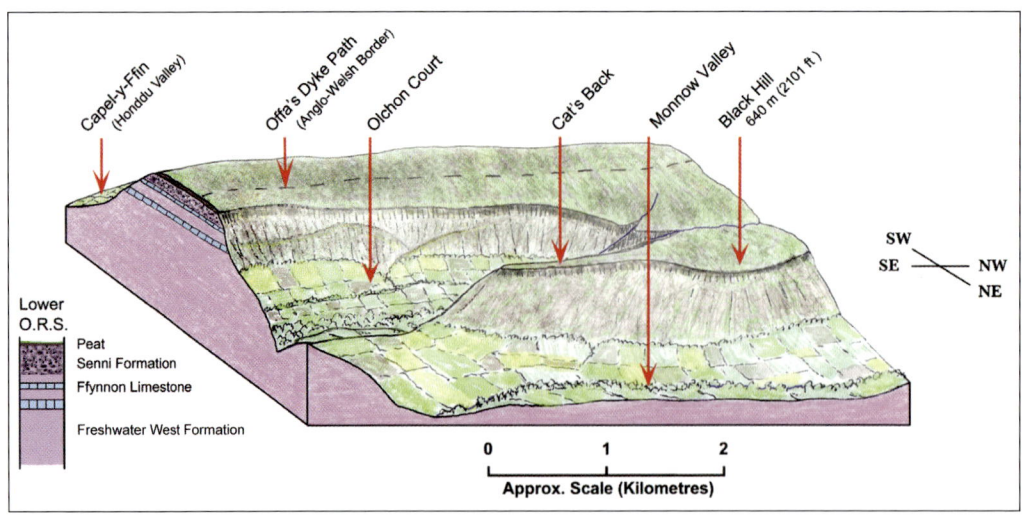

Figure 2.8 Block diagram of the Olchon Valley and Cat's Back. (Drawn by Peter Thomson with further work by Gerry Calderbank)

our county (Fig. 2.8). On the eastern side of this is the Cat's Back, a narrow ridge with spectacular views and most easily reached from a car park at its south-east end (Fig. 2.9).

From the Bradnor Hill car park, a short walk can be taken around the top of the hill by following the road for about a quarter of a mile and then bearing left along a track and continuing on a left-handed circular route back to the car park. (Take care not to interfere with golfers and watch out for flying golf-balls!).

Between Bradnor Hill and Hergest Ridge, the A44 brings the visitor into Herefordshire from Radnorshire – in particular the Vale of Radnor. The county boundary, about two miles to the west of Kington, is marked by the spectacular Stanner Rocks (Fig. 2.10) adjacent to the main road, which can be seen from the top of Bradnor Hill. Just on the Welsh side of the border, Stanner, together with two other hills, Worsell Wood and Hanter Hill, is

Figure 2.9 Members of the Woolhope Club's Geology Section on a visit to the Cat's Back in 2010. (© Robert Williams)

Figure 2.10 Stanner Rocks. (© John Stocks)

a Precambrian intrusion, comparable in age to the Malvern Hills in the east of the county. The three hills mark the line of the very ancient Church Stretton Fault, which runs from North Staffordshire via Church Stretton in Shropshire to south-west Wales.

On returning to the car park, look again at the Herefordshire Plain to the south. There, 20,000 years ago, a much more recent geological event was taking place, namely the last glaciation of the Quaternary Period. Today we are living in a warm inter-glacial period, between the last glacial period and the next prospective glaciation, albeit a warm period that we are exacerbating by human activity to create the 'global warming' now much spoken about. The viewer can perhaps imagine the scene. The great Wye Glacier pushed down the Wye Valley from the west (past Hay-on-Wye) and emerged into the northern part of the Herefordshire Plain. There it covered the entire Silurian/Devonian base, spreading east as far as Leominster and south as far as Hereford. It backed up against Bradnor Hill which, along with the 'cornstone' hills of the centre of Herefordshire and the peaks of the Black Mountains may have risen above the ice – 'nunataks' as geographers call such peaks above ice-sheets.

The presence of the ice affected the courses taken by pre-existing rivers, creating 'river diversions' (see View No. 3). The ice created erosive features in upland areas and depositional features on the lowland plain as it eventually melted about 12,000 years ago. There were also large areas, valleys between hills or areas which were dammed by 'glacial moraine', where lakes developed. A good example of the former is the beautiful Vale of Radnor beyond Stanner Rocks on the Welsh side of the border, clearly visible from Bradnor Hill. A good example of the latter is an area known today as 'Letton Lakes' between Bredwardine,

Letton, Norton Canon and Kinnersley, where water was impounded between the receding ice front and the ridge of moraine at Staunton on Wye (SO 37 45). Much of Letton Lakes is now an area of 'flood meadow', frequently still very wet in winter although used as grazing and hay meadows in summer. Herefordshire Wildlife Trust has acquired large parts of it as a Trust reserve (Sturts and Waterloo Reserve) to safeguard the distinctive flora and fauna, increasingly at risk from modern arable methods.

Box 2.2 **Roderick Impey Murchison** (1792-1871)

Roderick Murchison merits special mention in this book. Working in the 1830s, he was the first to establish a comprehensive and reasonably accurate geology of the Palaeozoic rocks of the Welsh Borders, including, of course, Herefordshire, and south Wales. He recorded this work in his book *The Silurian System* (1839), which laid the groundwork for much subsequent research and is still a valuable source of information.

Murchison was born in Scotland and served in the British army during the Napoleonic Wars. At the end of the war in 1815, he became interested in the new science of geology and, in association with other early 'pioneers' (especially the Revd Professor Adam Sedgwick, Woodwardian Professor of Geology in Cambridge), he began to study and map tracts of the British Isles in Wales and the southwest of England (to the west of those areas mapped by William Smith). In connection with the Welsh explorations, Murchison worked from 'the known into the unknown' – i.e. from England into Wales – while Sedgwick worked from 'the unknown towards the known' – i.e. from north-west Wales towards the Welsh/English border.

In 1831, Murchison undertook a tour of England and Wales with his wife, Charlotte, to make an investigation of the as yet unexplored older Palaeozoic rocks. It was during this tour that he first visited Herefordshire, meeting many local geological enthusiasts. Having stayed with a leading landed family in the Vale of Radnor, the Cornewall Lewises at Harpton Court – 'border squires' of some renown with an interest in geology – he was introduced to the Banks family of Kington and others with keen geological interests.[4] Important amongst these was the Revd T.T. Lewis, the curate of Aymestrey (Box 2.4). Murchison was to walk on Hergest Ridge and Bradnor Hill with his local friends, examining its quarries and searching out distinctive fossils. A (private) collection of geological finds from that period is still maintained by the Banks family.

In 1839 his masterly work *The Silurian System* was published, describing the geology of the border area of England and Wales, including Herefordshire.[5] The text was accompanied by magnificent coloured maps, cross-sections,

Figure 2.11 Murchison in 1836, at the time of his work locally. (©Trustees of the British Museum)

landscape drawings (some done by Charlotte) and detailed engravings of fossils. (An example cross-section, although not from this book, is shown in Fig. 2.21.) Many of the subscribers were Murchison's local friends in the Marches area. They included the Banks family in Kington and the Revd T.T. Lewis (Box 2.4), who was a knowledgeable geologist, having himself been a student of Sedgwick at Cambridge. A wider view of the Silurian rocks was published as *Siluria* in 1854. When the Woolhope Naturalists' Field Club was established in 1851, Murchison and other distinguished national geologists became honorary members. Murchison led at least one field meeting for the Club, on the Malvern Hills. He later did extensive field work in Russia. In 1855 he was appointed director-general of the British Geological Survey.

Murchison's work in Herefordshire is of particular interest in this book and is mentioned in later chapters. He was clearly aided in this work by his contacts with local geological enthusiasts, many of whom had performed significant work in their own limited areas and shared their findings with him.

In *The Silurian System* there is an illustration of what is now the A44 in Herefordshire, a road which Murchison must have used many times. This is a fine drawing in the Picturesque style by Joseph Murray Ince (1806-1859)[6] depicting Stanner Rocks, Worsell Wood and Hanter Hill, the line of the Church Stretton Fault in this area, and Hergest Ridge (Fig. 2.12).

For biographies of Roderick Murchison see, for example, Sir A. Geikie's *Life of Sir Roderick I. Murchison* (1875) and J.L. Morton's *King of Siluria: How Roderick Murchison Changed the Face of Geology* (2004).

Figure 2.12 The A44 and Church Stretton Fault: illustration in The Silurian System, *by Joseph Murray Ince*

View No. 3: *Croft Ambrey*: approaching Herefordshire from the North

Two main roads enter Herefordshire from the north (the A4110 and the A49, which passes through Leominster). North of Leominster and between these two roads a line of hills rises from the Herefordshire Plain to the north of a minor road (B4362). Croft Ambrey is an Iron Age hillfort (SO 445 668) on this line of hills, north of Croft Castle, a National Trust property just off the minor road. There is a car park at the castle and an easy walk to the top of the hill – about a mile and a half through National Trust and Forestry Commission land. [OS Landranger 137 'Ludlow & Church Stretton' and 149 'Hereford & Leominster']

From the summit of Croft Ambrey, there are spectacular views in all directions, with particularly striking views to the Black Mountains and the Malvern Hills, on the western and eastern boundaries of the county,

Southward the view is to Leominster, Dinmore Hill and Hereford beyond; while, to the north, it is across the Vale of Wigmore to Shropshire. On a clear day, Titterstone Clee hill can be seen to the north-east beyond Ludlow. The hill has a small remnant of Carboniferous rocks at the summit (Fig. 2.14). Although in Shropshire, this landscape feature is linked geologically with the only Carboniferous outcrops in south Herefordshire in an area known as The Doward, both having once been on the southern side of the one-time Carboniferous land mass of St George's Land.[7] St George's Land stretched east-west across Wales and the English Midlands, separating the areas of marine deposition to the north and south. (See View No. 4 and Chapter 8).

Croft Ambrey is on the north-western ridge of Silurian rocks which emerge from below the Herefordshire Plain, forming a line of hills between Ludlow and Kington – Bircher

Figure 2.13 View south-east from Croft Ambrey: the Malvern Hills on the centre of the skyline. (© Robert Williams)

Figure 2.14 View to the north-east from Croft Ambrey towards Ludlow and Titterstone Clee. (© John Stocks)

Common, Croft Ambrey, Shobdon Hill, Wapley Hill, Bradnor Hill and Hergest Ridge. This ridge is also the southern flank of a great fold of rocks lying to the north, the Ludlow Anticline, which has its northern flank between Ludlow and Leintwardine, an area sometimes known as The Vale of Wigmore (Fig. 2.15).

The valley to the immediate north of Croft Ambrey follows a geological fault running north-east to south-west in the southern flank of the anticline so that the rocks of the Croft Ambrey ridge are repeated in the hill with the large working quarry at Leinthall Earls, clearly visible on the north side of the valley (Figs. 2.15 and 2.16). Croft Ambrey is seen to lie on a ridge of Aymestry Limestone and it is this same rock which is worked in the quarry.

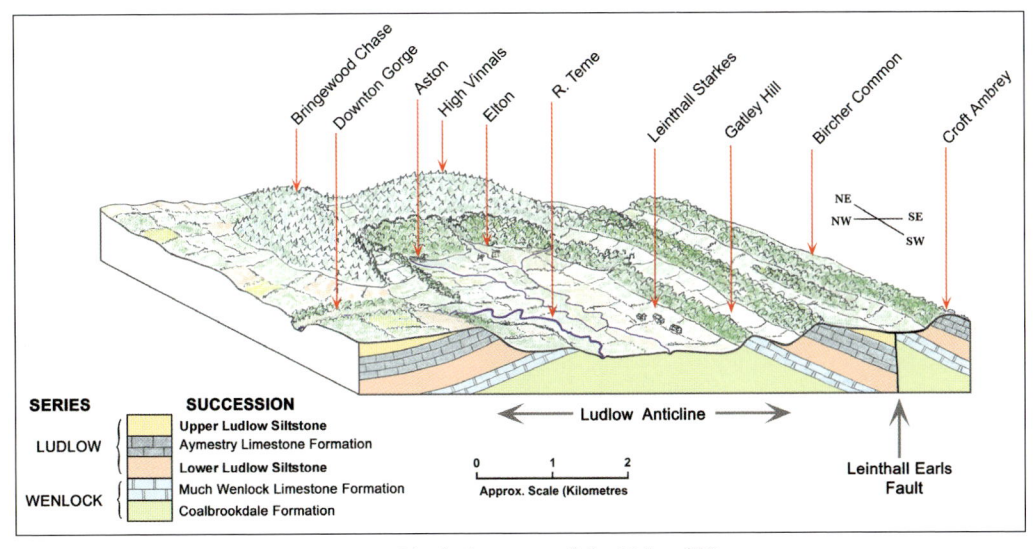

Figure 2.15 Block diagram of the Vale of Wigmore. (Drawn by Peter Thomson with further work by Gerry Calderbank)

Figure 2.16 Leinthall Earls Quarry. (© John Stocks)

Beyond the quarry can be seen the flatter land of the Vale of Wigmore, formed by the erosion of the folded rock over millions of years. The older rocks towards the centre of the fold include the Much Wenlock Limestone, of which Wenlock Edge in Shropshire is a famous and distinctive feature.

During the last glaciation (about 20,000 years ago), the Vale of Wigmore (i.e. the interior of the eroded fold) filled with water and became a lake as the Wye Glacier extended across north Herefordshire. The glacier blocked the River Teme near Aymestrey, preventing it from following its earlier southward course from Leintwardine to join the River Lugg (and the Wye river system). As a result, the Teme now flows into the Vale of Wigmore at Leintwardine and out through a gorge at Downton (Fig. 2.17), probably cut by lake overflow, on the north side of the Vale, flowing then north-east to Ludlow and becoming eventually a tributary of the River Severn at Worcester.[8]

Figure 2.17 Downton Gorge and the River Teme. (© Peter Wakely/Natural England)

Box 2.3 **Archaeological Sites**

As an Iron Age hillfort, Croft Ambrey is similar in age and construction to the British Camp on the Herefordshire Beacon (see View No. 1). In this case, however, it is constructed on a ridge of Silurian rather than Precambrian rocks – a difference in geological age of about 300 million years. Hillforts such as these were to some extent replaced by Roman towns as centres of settlement after the Roman occupation of Britain in 43 A.D. (e.g. *Bravonium*, today's Leintwardine, on the River Teme). In the years following the departure of the Romans (410 A.D.) many were re-occupied by a new generation of people as part of the defence of their, by now, Romano-British Christian culture against the invading Saxons. One early chronicle speaks of a leader of such defences called Ambrosius, who successfully held the line for a time; and it is from him that some early antiquaries thought the name 'Ambrey' may be derived.[9] Over the years, oral tales of such heroism transformed this Roman-British Christian warrior into a medieval knight and a thousand years later Ambrosius became Arthur in Mallory's famous poem *Le Morte d'Arthur* (15th century)!

Other well-known Iron Age hillforts in the county are British Camp and Midsummer Hill on the Malvern Hills and Credenhill near Hereford.

In Herefordshire there are at least two distinctive references to Arthur with geological overtones: (1) Arthur's Stone (SO 3188 4313), a Neolithic burial chamber of Old Red Sandstone in the western side of the county above the River Wye at Bredwardine and (2) King Arthur's Cave, a Palaeolithic site on The Doward in the south of the county (see View No. 4) where there is a large cave in the Carboniferous limestone cliffs above the River Wye (SO 5458 1558). (See Fig. 2.25)

The Saxons eventually drove the Romano-British Christian people westward, where their traditions became the core of Welsh culture and language. In central England, including Herefordshire, the Anglo-Saxon kingdom of Mercia was then established. One of its kings, Offa, created the famous Offa's Dyke from north to south in this 'Marches' borderland as a demarcation line between the two peoples; and today a long-distance trail has been established which follows this route, part of which crosses Bradnor Hill (see View No. 2) – an ideal means to geological and landscape observation.

Much more information on archaeological sites in Herefordshire may be found in Keith Ray's *The Archaeology of Herefordshire: An Exploration* (2015).

Figure 2.18 Arthur's Stone. (© John Stocks) (See Fig. 2.25 for King Arthur's Cave)

Box 2.4 **Thomas Taylor Lewis (1801-1858)**

Figure 2.19 Aymestrey Church and Pokus Wood. (© Robert Williams)

The rocks of northern Herefordshire became of great interest to early geologists in the 19th century largely as a result of the work of the Revd Thomas Taylor Lewis, the one-time curate at Aymestrey, a village near to Croft Ambrey on the River Lugg. Lewis was one of a long and honourable line of 'English Parson Naturalists',[10] who, in addition to their parochial responsibilities, took a delight and interest in the natural world around them. Lewis's special interest was geology. His family had had coal mining interests on Titterstone Clee Hill and he himself had been a student of the Revd Professor Adam Sedgwick at Cambridge.[11] In July 1831 he was visited by Roderick Murchison, who was travelling through Radnorshire into Herefordshire and had heard of him through an acquaintance in Kington.

As a result of his walks with Lewis on the hills around Aymestrey, Lucton, Croft Ambrey, Wigmore, Leintwardine and Downton, Murchison was eventually, in *The Silurian System* (1839), to name 'the central member of the Ludlow rocks' after the village of Aymestrey. The rock's name, 'Aymestry (*sic*) Limestone', thus passed into international geological parlance, where it still remains (Fig. 6.28).[12] Indeed, Murchison somewhat touchingly expresses the hope in *The Silurian System* that 'The application of his leisure hours to the cultivation of the natural history of his neighbourhood may one day enable Mr. Lewis to confer upon Aymestry the celebrity which White has bequeathed to Selborne.' (cf. The Revd Gilbert White *The Natural History and Antiquities of Selborne* [1789])

In the event Lewis did not write a 'Natural History of Aymestrey' but, by naming one of the Silurian limestones in the Ludlow series of rocks after Aymestrey, Murchison ensured that the village and the surrounding area would take on an international significance in geological circles.

Later, Lewis, by then vicar of Bridstow near Ross-on-Wye, became the second President of the Woolhope Club.

More details about T.T. Lewis and other early members of the Woolhope Naturalists' Field Club may be found in the chapter by J.H. Ross in *A Herefordshire Miscellany* (2000) and K. Andrew 'History of Geology in Herefordshire' (to be published).

Figure 2.20 T.T. Lewis as the curate at Aymestrey

This gorge was created as a sub-glacial channel in a similar fashion and at around the same time as the better known Ironbridge Gorge on the River Severn in Shropshire.

One of the distinctive marker fossils for the Aymestry Limestone is the brachiopod *Kirkidium Knighti* (see Fig. 6.1), named by Murchison after Richard Payne Knight's brother, Thomas Andrew Knight (1758-1838), who lived at Elton Hall in the Vale of Wigmore and later at Downton Castle, and with whom Murchison stayed during his travels in this area. Thomas Andrew Knight was a distinguished horticulturalist, a Fellow of the Royal Society and one of the founders of the London Horticultural Society (LHS) in 1804, which became the Royal Horticultural Society (RHS) in 1864. From 1811 until his death in 1838, Thomas Andrew Knight was President of the LHS.

In the 1850s, Murchison devised a splendid 'visual aid' for a talk he gave in Ludlow on the Ludlow Anticline (Fig. 2.21). It very clearly shows a section through the Ludlow Anticline. Given the pioneering work that had only just been done on the local geology and geomorphology, this is a remarkable achievement. The original, some ten feet long, is in Ludlow Museum.

Figure 2.21 Murchison's cross-section of the Ludlow Anticline (lower picture). (The upper picture shows his understanding of the Palaeozoic sequence from the Carboniferous, through the Devonian and Silurian to the Cambrian). (© Shropshire County Museum Service)

View No. 4: *Coppet Hill*: approaching Herefordshire from the South

The main entrance to Herefordshire from the south is the A40 following the course of the River Wye upstream from Monmouth to Ross-on-Wye. South of Ross turn right, take the B4229 to Goodrich (SO 575 195). After about 1.5 miles fork left off the B road into the village and park in the centre near the shop or follow the signs to Goodrich Castle and park in the pay car park by the visitor centre. Return on foot to the village centre turning left onto the minor road south, crossing over the B4229 on a bridge, and follow the road uphill to SO 577 188 where the road forks. Then follow the footpath leading off between the two roads, up steps towards Coppet Hill Common. The footpath passes a fine exposure of Huntsham Hill Conglomerate (was called Quartz Conglomerate; Chapter 7) on the way up. [OS Landranger 162 'Gloucester & Forest of Dean']

From Coppet Hill there are magnificent views of the valley of the River Wye as it flows south from Ross-on-Wye beneath Chase Hill to Goodrich, sweeping in large meanders under Yat Rock (in Gloucestershire, east of the river) past Huntsham Hill to enter the Wye

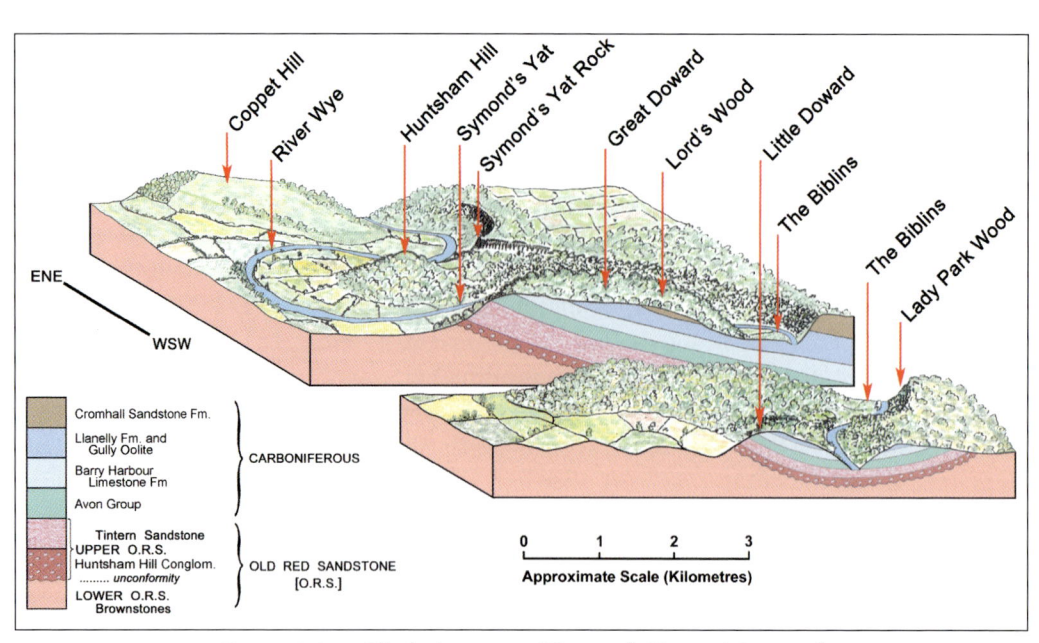

Figure 2.22 Block diagram of Symonds Yat and Doward.
(Drawn by Peter Thomson with further work by Gerry Calderbank)

Gorge at Symonds Yat. North of Chase Hill is the undulating skyline of the Woolhope Hills, an anticline of Silurian rocks. On a clear day the asymmetric outline of Titterstone Clee hill in south Shropshire can be seen beyond the west end of the Woolhope Hills. Due north the skyline is formed by Dinmore Hill, then interrupted by Aconbury Hill, the northernmost outcrop of the Brownstone rocks of the Lower Devonian. To the west are the Black Mountains forming the western boundary of the county and to the south-west are splendid views to the Great Doward. Having surveyed the scene, a supplementary visit to the Dowards is recommended, in particular Little Doward and the Wye Gorge.

The Dowards are distinctive as the locality for Herefordshire's only rocks of the Carboniferous Period with their associated landscape features. The Carboniferous Period succeeded the Devonian, its deposits of limestone being laid down in tropical conditions when this region was close to the Equator, and the area has long attracted geologists, naturalists and lovers of the 'picturesque' and 'romantic'. In the area of Great Doward (SO 55 16), there are eight reserves of the Herefordshire Wildlife Trust, the ecological interest of these varying from the remains of the primary woodland to secondary woodland on old pasture, old quarries and the development of grassland on filled quarries. This is a measure of how important the area is from a natural history point of view – that being to a large degree influenced by the local geology.

The Carboniferous rocks of the Dowards lie above the older Silurian/Devonian rocks, which in their various geological strata form the central part of Herefordshire. Here, in the south of the county, the beds dip southward with the River Wye flowing across the area from north to south. The north-facing escarpments allow for some dramatic views into the county, the most famous and distinctive viewpoint being Yat Rock on the left bank of the Wye as it cuts a meander and forms a cliff in the Carboniferous Limestone.

Figure 2.23 Yat Rock from Coppet Hill. (© Robert Williams)

Standing at Yat Rock (SO 564 160) and looking north (Fig. 2.24), one sees the Wye with Coppet Hill on the right bank to the north-east. Coppet Hill is formed of very strong conglomerate rock with sandstone and outcrops of the Carboniferous limestone.

Figure 2.24 View from Yat Rock to Coppet Hill. (© Robert Williams)

Follow the river upstream and the visitor comes to Goodrich Castle and eventually Ross-on-Wye, both on distinctive Brownstone rocks of the Old Red Sandstone.

Before the Pleistocene Epoch of the Quaternary Period (about 2.5 million years ago), the River Wye flowed across a wide plain, forming great sweeping meanders. Since then falling sea-levels have caused the river to cut more vigorously into the underlying strata, thereby creating the 'incised meander' (i.e. one cutting down into the rocks) and giving us the gorge which we see today (Chapter 9). The falling sea-level, however, has not been continuous and, during pauses in the process, river terraces of sediment (sands and gravels) built up – these now being left above the present flood plain in various parts of the Wye Valley.[13]

To get to the Dowards, take the B4229 south from Goodrich to the A40 and turn south towards Monmouth. Turn left after about a mile at Whitchurch (Symonds Yat West turning), turn right at the roundabout and then first left for Crockers Ash. From there, follow the signs for Doward Park Campsite and the Biblins. At the camp site (SO 548 157), turn south onto the track and after about 100m there is a parking area just before the Forestry Commission sign. (Take care on the narrow roads in this area of the Dowards: it is a popular place for visitors and there is also the normal residential traffic.)

To explore the area of the Dowards, the EHT Trail Guide entitled *The Wye Gorge*[14] provides an excellent introduction to the Carboniferous deposits of this part of the county. The trail is, however, 'best suited to experienced walkers and those able to read maps. Some sections are steep and uneven and can be slippery. Stout footwear is recommended. The route will take approximately three to four hours.' As well as providing a fine close-up view of the Wye as it flows through the gorge, this trail includes among its many sites King Arthur's Cave (SO 5458 1558; see also Chapter 9) in the Carboniferous rocks.

Figure 2.25 King Arthur's Cave. Evidence of Palaeolithic animals and human occupation has been found here. (© John Stocks)

| Box 2.5 | **Geology and Artistic Movements** |

The development of geology as a science was contemporary with a growing appreciation of the aesthetic of the 'landscape'. In the region of the lower Wye particularly, the early exponents of the 'Picturesque' and the 'Romantic' began to explore the beauties of the craggy rocks, cascading water and the effects of both on the natural vegetation. The Revd William Gilpin (1724-1804) publicised the beauties of the Wye Gorge; while Admiral Nelson and Lady Hamilton came 'on excursion' down the river from Ross-on-Wye. William Wordsworth too travelled in this area, famously to Tintern in Monmouthshire where he wrote *Lines Composed above Tintern Abbey*, but also further north, visiting his wife's relations who farmed in the Vale of Radnor and later in north-west Herefordshire. Indeed, the Picturesque and the Romantic Movements have had an enduring influence not only on our understanding of landscape but also more generally on our understanding of the natural order. Combined with 18th and 19th century scientific insights, this aesthetic has contributed to the development of contemporary concerns about conservation; and, as already mentioned, this area has a number of Nature Reserves, the flora of many of the reserves reflecting the underlying geology. There are some excellent guides to these, obtainable from Herefordshire Wildlife Trust. (The Picturesque Movement is further discussed in Box 10.1.)

View No. 5: *Woolhope Dome*: a more central viewpoint

This distinctive upland area, lying between Hereford and Ledbury, can be approached from either town along the A438 in the north, from the A449 between Ledbury and Ross-on-Wye in the south-east or from the B4224 between Hereford and Ross-on-Wye in the west. The village of Woolhope (SO 612 357) lies at the centre of this feature and visitors are advised to explore views looking inward as well as outward from this geological/geomorphological formation. Several positions are recommended (see below). (Be careful on the narrow minor roads that encircle and cross the area.)

As a geological feature, the Woolhope Dome is an area of strongly folded older Silurian rocks which have been revealed by erosion of the younger Old Red Sandstone rocks of the central part of Herefordshire.[15] Views inward are therefore towards the older rocks. Here, there are parallels with the view to the north from Croft Ambrey, described earlier, the Ludlow Anticline being of the same age as the Woolhope Dome, formed of the same rocks and with the weaker layers (siltstones and shales) of the dome-like fold being now eroded away leaving the stronger limestone layers as upstanding ridges (Fig. 2.26). In consequence, the outer flanks of the Woolhope structure dip under the Herefordshire plain, while the inner slopes are 'escarpments' left by the differential erosion of the upfolded rocks.

On the outside is a ridge of the youngest rocks, the Aymestry Limestone. Closer in is a ridge of the older Much Wenlock Limestone and beyond that again towards the centre at Woolhope is the oldest Woolhope Limestone. Just to the north-west of Woolhope is Haugh (pronounced 'Hoff') Wood. The underlying rocks here are sandstones and siltstones and, being at the centre of the dome, are the oldest in the area. On a much larger scale, similar

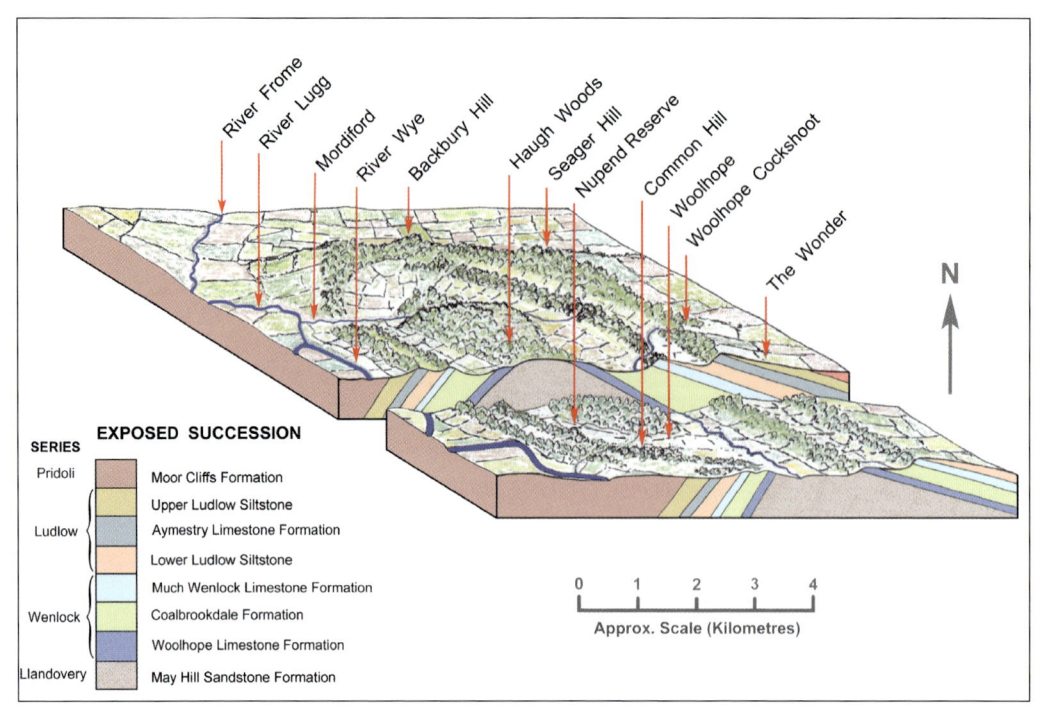

Figure 2.26 Block diagram of the Woolhope Hills. Because of the erosion of the circular dome-like upfolded rocks, the geological map shows an area of concentric 'circles', each representing the exposure of different types of rock.
(Drawn by Peter Thomson with further work by Gerry Calderbank)

eroded 'domes' or anticlines, with inward-facing escarpments, exist in the Weald of south-east England and in the Derbyshire Peak District. Haugh Wood is owned by the Forestry Commission.

A good way of exploring this area is to cross it from east to west, seeing views both outward to the surrounding parts of Herefordshire and inward to the centre of the 'Dome'. At the same time, the traveller crosses the different outcrops of rock, gaining a good idea of the topography of the inner landscape. To start, about half a mile south-west of the crossroads and traffic lights at Trumpet, turn south (SO 648 391) off the A438 towards Putley Common. At Woolhope Cockshoot (SO 631 372) turn north-west and take the road to Checkley (SO 599 384) and Prior's Frome (SO 576 390). The road to Checkley follows the outcrop, in a valley, of the Lower Ludlow Siltstone before passing through the escarpment of the Much Wenlock Limestone via a faulted gap (SO 612 380) eroded by the Pentaloe Brook. It then follows another valley in the outcrop of the siltstones of the Coalbrookdale Formation with Haugh Wood to the south and the peak of Backbury Hill with its Iron Age hill fort on the Aymestry Limestone dominating the skyline ahead. Before reaching Prior's Frome there is a car park and picnic site on the left at Swardon Quarry (SO 578 385), now disused. This quarry exposes a limestone within the Upper Ludlow Siltstone. Beneath projecting rocks at the top of the exposure is a layer of bentonite, a volcanic ash laid down as the limestone was being formed.[16]

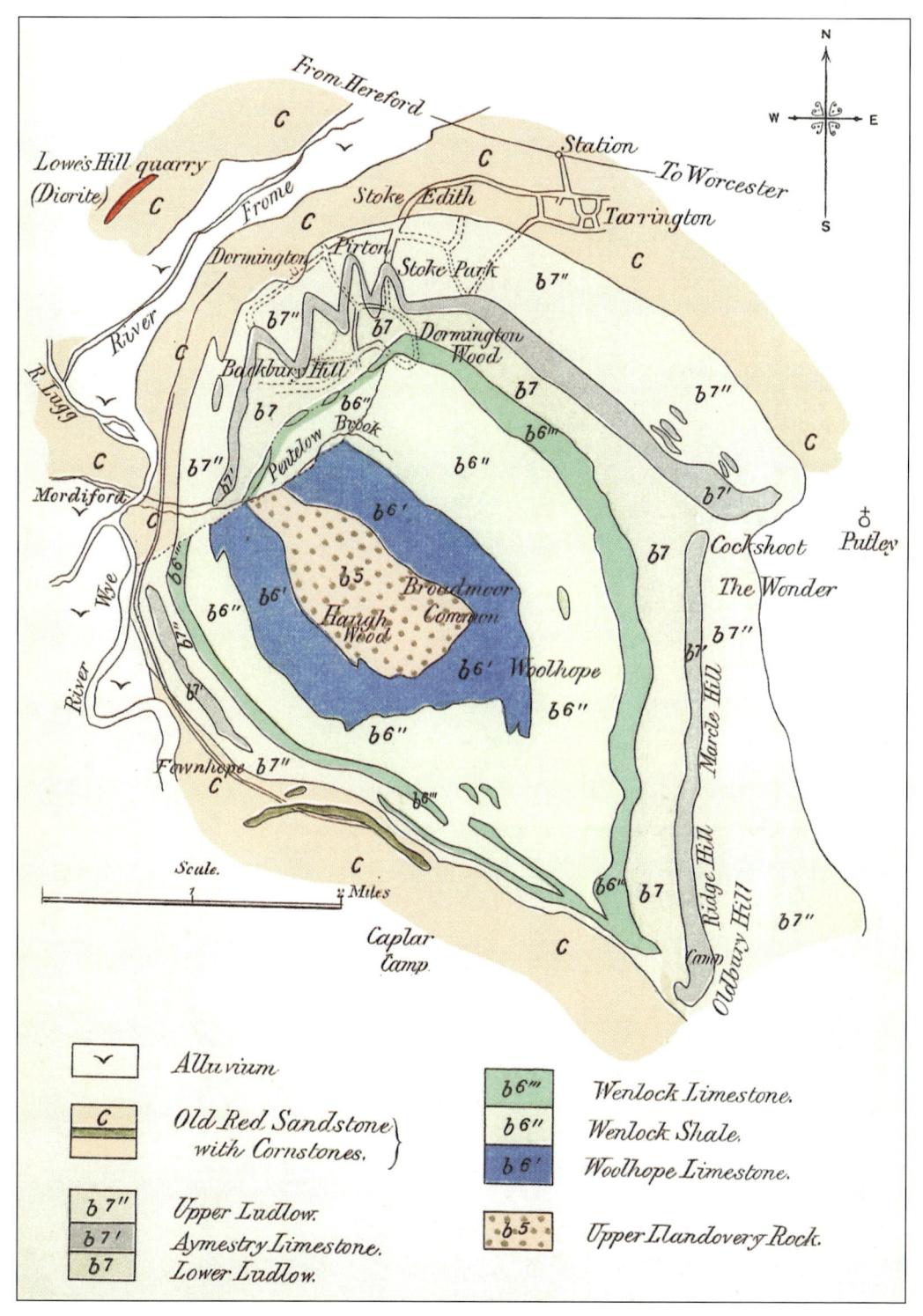

Figure 2.27 Geology map of the 'Woolhope Dome', after which the Woolhope Naturalists' Field Club was named. (G.H. Piper, 1891 Transactions of the Woolhope Club, pp.164-168)

Figure 2.28 (above) View of the centre of the Woolhope Dome, looking north-east towards the oldest rocks at Haugh Wood from the Wye Valley Walk on Common Hill (SO 588 347). (© John Stocks)

Figure 2.29 (below) View of the Woolhope Dome from the north-west, near Bartestree (SO 566 409). (© John Stocks)

Park here and climb the steps on the south side to the picnic site. From here a fine westerly view can be obtained towards Hereford city sited on the outwash gravels of the last glaciation, surrounded by the 'cornstone' hills of central Herefordshire. Beyond them, on the south-west horizon, are the sandstones of the Black Mountains. To the north are the hills of north-west Herefordshire, with Silurian rocks similar to those in this area.

Return to Woolhope Cockshoot and turn right following the road to Winslow Mill (SO 625 361). Turn left (signposted to Much Marcle) and follow the road along the ridge of the Aymestry Limestone on Marcle Hill. At the first road junction there is a car park (SO 631 347), in a disused quarry of Aymestry Limestone, at Sleaves Oak.

From the public footpath at the road junction there is a fine view eastwards towards Ledbury and the Precambrian rocks of the Malvern Hills. Beyond Ledbury are Silurian rocks like those of the Woolhope Dome. Between this viewpoint and Ledbury are the younger Old Red Sandstone rocks with the 'cornstone' Wall Hills rising above them west of Ledbury, in the valley of the River Leadon. South-east of the Malverns the Cotswold Hills, of oolitic limestone, appear on the horizon on the other side of the Severn Valley.

Now retrace your route to Winslow Mill, turn left and follow the road westward via Woolhope to Fownhope on the B4224.

At Winslow Mill the road follows a stream cutting through the escarpment of the Much Wenlock Limestone and continues across the vale (in siltstones of the Coalbrookdale Formation), climbing up into Woolhope village, sited on the Woolhope Limestone, before dropping down again onto the Coalbrookdale Formation.

Just west of Woolhope, the central part of the Dome at Haugh Wood can be seen to the north through a field gate (SO 592 352).

Continue on the road to Fownhope. Finally the road passes through a deep gorge through the Much Wenlock and Aymestry Limestones to reach Fownhope (SO 577 345).

On reaching Fownhope you will have crossed the Woolhope Dome! At Fownhope, you are close to the River Wye, here on its way south towards Ross-on-Wye and View No. 4 near Coppet Hill. Turn right and follow the B4224 north. Between the next T-junction and the village of Mordiford (SO 572 374), the Wye is joined by the River Lugg, which has now crossed Herefordshire from the north via Aymestrey and Leominster. From Mordiford, take the road north to Dormington, back to the A438. On reaching the main road, turn left (SO 584 403) for Bartestree, about a mile along the road. About 400 yards south of the crossroads in Bartestree, park at the junction with a minor road on the left and walk a few yards to the reference point (SO 566 409). Here an external view of the Dome may be obtained, a view now from the north-west towards Backbury Hill on the Aymestry Limestone ridge.

And one final point: as indicated earlier in this chapter, the Woolhope Dome allows an insight into not only the geology and landscape of the area, but, in a rather particular way, something of the history of geology in Herefordshire and beyond. This was the geological formation and landscape from which the Woolhope Naturalists' Field Club took its name in 1851. It was one of many Victorian scientific natural history societies which grew up across the country in the 19th century. The early honorary membership included leading national figures in the science of geology: Roderick Murchison; Revd Professor Adam Sedgwick, Woodwardian Professor of Geology at Cambridge; and Sir Charles Lyell, the author of *Principles of Geology* (1830-1833) which Charles Darwin took with him on his *Beagle* voyage (1831-1836) and which so influenced him in his reflections on 'natural selection'.

The early club members were particularly fascinated by the local geology of the Woolhope Dome and especially the fossiliferous nature of the Silurian deposits. They were somewhat disappointed by the sparsity of Devonian fossils in this area, but the abundance of Silurian rocks and fossils of the Woolhope Dome particularly appealed to them. For many enthusiasts in the new science of geology, the name Woolhope (as of Aymestrey) was as central to their thinking as the 'Dome' was to the local landscape of Herefordshire. (See EHT Trail Guide: *The Woolhope Dome*).

Further reading
Geology
Allott, Andrew *Marches* – number 118 in the Harper Collins New Naturalist series (2012)
Hutton, James *Theory of the Earth with Proofs and Illustrations,* (1795)
Lyell, Charles *Principles of Geology,* (1830)
Murchison, Roderick *Siluria,* (1854)
Toghill, Peter *The Geology of Britain: An Introduction,* Swan Hill (2000)
H&W Earth Heritage Trust *Explore Geology and Landscape Trail Guides for Woolhope Dome. Malvern Hills 1, Kington & Hergest, Wigmore Glacial Lake and Wye Gorge*
Jenkinson, Andrew *Mortimer Forest geology trail,* (1991), 24pp. (Scenesetters), E-mail: mail@scenesetters.co.uk. (Ludlow: Forestry Commission)
Peterken, George *Wye Valley* – number 105 in the Harper Collins New Naturalist series (2008)
Thompson, Mike *Coppet Hill and Goodrich: Walks from Goodrich Castle*
'A Picnic in Siluria', Woolhope Club (2008), produced by Field of Vision – A DVD of a loose re-enactment of a mid-19th century field trip of the Woolhope Club
British Regional Geology: The Welsh Borderland, HMSO (1971)

Archaeology
Ray, Keith *The Archaeology of Herefordshire: An Exploration,* Logaston Press (2015)

The Picturesque Movement
Appleton, Jay 'Some thoughts on the geology of the picturesque' in *Journal of Garden History,* no.3, (1986), pp.270-291
Daniels, S. and Watkins, C. (eds), *The Picturesque Landscape: Visions of Georgian Herefordshire,* Dept. of Geography, University of Nottingham in association with Hereford City Art Gallery and University Art Gallery, Nottingham (1994)
Gilpin, William *Observations on the River Wye and Several Parts of South Wales,* (1782) (https://archive.org/details/observationsonr00gilpgoog)
Price, Uvedale *Essays on the Picturesque,* (1810) (https://archive.org/details/essaysonpictures01priciala)
Watkins, C. and Cowell, A. *Uvedale Price (1747-1829); Decoding the Picturesque,* The Boydell Press (2012)

3 Basic Geology

Stand astride the backbone of the Malvern Hills and contrast the landscape to the east, the plain of the River Severn, with the rolling hills of Herefordshire to the west. These very different landscapes are the results of various geological processes, which have taken place over many millions of years at various periods of the Earth's history. A major aim of geological science is to describe these processes and how they have given rise to the landscapes we see today. This is the subject of Chapters 4 to 9 of this book. The present chapter outlines the basic geological concepts.

Time

The Earth has an age of some 4,600 million years. This figure has been reasonably well established since the middle of the 20th century using scientific techniques which were not available to the early geologists at the start of the 19th century. A major goal of these first geologists was to reconcile their observations with the biblical account of the Earth's origin. This proved to be difficult from the beginning and it took about two centuries before the currently accepted figure was determined in the mid-20th century.

One of the most striking insights of the early geologists was to recognise that the timescale which had been used up to their day was inadequate to account for the age of the rocks as they were coming to understand them. The vast depths of sediment in, for example, the Black Mountains and the Brecon Beacons gave pause for thought. As a solution to this problem, the early geologists gradually developed the idea that earth processes observable today could help to explain those which had formed rocks and geological features in the past. This concept was called 'uniformitarianism' – the idea that 'the present is the key to the past' – a theory developed by James Hutton (1726-1797) and particularly by Sir Charles Lyell (1797-1875). Knowing the observed processes of erosion and deposition (as after a heavy storm), it was clear that familiar landforms must have taken millions, not thousands of years to be fashioned. Specific methods of dating were not available to the first geologists, but by the early 19th century most people could recognise that Archbishop Ussher's careful calculation from the biblical text in the 17th century that the world had been created on 22 October 4004 B.C. (on a Saturday at about 6pm!) did not quite match what they observed.

Relative and Absolute Dating

For many years geologists had no way of establishing the age of rocks; they could only postulate the *relative* age of one stratum compared to another. Stratigraphy, the study of

- GEOLOGICAL TIME -

Geologists work with a vast time scale, reaching back from the end of the last glaciation to the very beginnings of earth history. This spans 4,660 million years and therefore requires a sophisticated classification system.

The result is a range of terms, taken mainly from Greek and Latin, that has gradually evolved into the following hierarchy:

EON This fundamental unit derives from the Greek word 'aeon', meaning *"life, an age, or eternity"*. It has been defined as 10 to the power of 9 - and so the age of the earth is 4.66 eons.

ERA There are three such main (first-order) units. Geologists used to include the Quaternary but this is now classified as the uppermost 'Period'.

(The name is retained from when the earliest workers spoke vaguely of "Primary, Secondary, Tertiary and Quaternary rocks")

The current nomenclature refers to evolutionary stages in the development of life - separated by mass extinctions.

Period A second-order chronometric division, also known as a 'System' in chronostratigraphy when referring to its actual rock content and sequence.

In this case the nomenclature is varied in origin - including place names from Russia, Germany and the UK - plus two Celtic tribes and the usual Classics.

Epoch Epochs are chronometric (third-order) sub-divisions of Periods but are not shown on this diagram. For example, the Quaternary contains two Epochs, the Neogene has two, and the Paleogene consists of three Epochs.

In turn, their chronostratigraphic equivalents are called **Series**, further divided into (fourth-order) **Stages** and then **Chronozones**. Reverting to chronometry, the smallest unit of geological time is the **Chron**.

PHANEROZOIC EON

- TIME SCALES -
Non-linear & disproportional

ERA — Period / Duration

CAINOZOIC (Recent Life)
- Tertiary
  - 20.5 — Holocene ... (The present)
  - Quaternary — 2.58 Ma
  - Neogene — 23 Ma
  - 43 — Paleogene
  - 66 Ma ... MASS EXTINCTION

MESOZOIC (Middle Life)
- 79 — Cretaceous
- 145 Ma
- 56.3 — Jurassic
- 201.3 Ma
- 50.6 — Triassic
- 251.9 Ma ... MASS EXTINCTION

PALAEOZOIC (Early Life)
- Permian — 298.9 Ma
- 47 — Carboniferous — 358.9 Ma
- 60 — Devonian — 419.2 Ma
- 60.3 — Silurian — 443.8 Ma
- 24.6 — Ordovician — 485.4 Ma
- 41.6 — Cambrian — 541 Ma
- 55.6

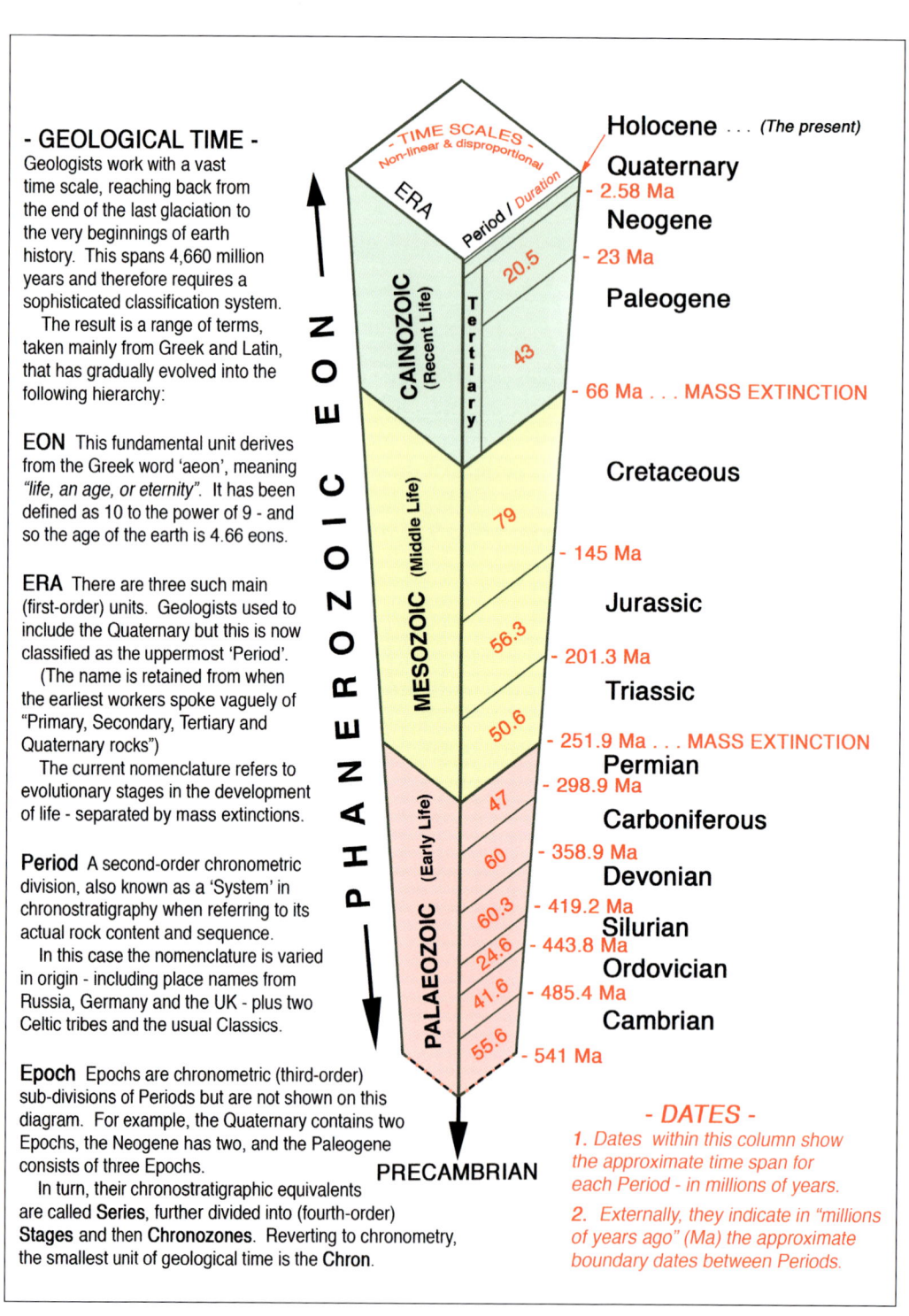

PRECAMBRIAN

- DATES -

1. Dates within this column show the approximate time span for each Period - in millions of years.
2. Externally, they indicate in "millions of years ago" (Ma) the approximate boundary dates between Periods.

Figure 3.1 The Geological Column.
(Drawn by Gerry Calderbank)

the sequence and composition of rocks, is one way of achieving this; another method is by comparing fossils found in different rock strata. Trilobites for instance, which are found in rocks of the early to mid-Palaeozoic Era and which displayed different physical characteristics as they evolved over time, are not found in rocks later than those of the Permian Period. In the last hundred years, however, scientific methods for the *absolute* dating of rocks – 'geochronology' – have been developed. Radiometric dating involves measurement of the radioactive decay of isotopes of certain chemical elements while fission-track dating exploits certain properties of the mineral zircon. The earliest rocks which have been found so far on the Earth are in northern Canada and are about 4,100 million years old.

By such means, chronological sequences of geological deposits could be given firm dates. Early geologists had named specific geological periods and the order in which they succeeded one another; now a true 'time scale' could be established.

The Geological Timescale

From their observations of rock types and the fossil record embedded within them, early geologists were able to establish and name distinctive geological 'systems', or 'periods'. These they often named after a place or local idea and, by observation, they were gradually able to arrange their systems in a chronological sequence. Today the names of those systems persist and these are identified with specific geological periods – Cambrian (from the Latinised form of 'Cymru', Wales), Ordovician (named after an iron-age tribe in west Wales), Silurian (after another Welsh iron-age tribe), Devonian (named after Devon), Carboniferous (descriptive of the carbon or coal measures) and Permian (from the city of Perm in Russia).

The division of geological time into Eons, Eras, Periods (or Systems) is shown in Figure 3.1 together with the currently accepted dates of their beginnings and endings (in millions of years before the present (Ma)). The various boundaries are chosen to coincide with major changes in the geological record; often this is a mass extinction. Each period is split into a succession of finer sub-divisions which correspond to smaller but widespread geological changes.

The Earth

The planets of the solar system grew out of a rotating disc of dust particles orbiting the star we now know as the sun. Particles aggregated within this disc to form larger and larger bodies which, colliding, breaking up and joining one another, finally formed the planets of the solar system, one of which was the Earth. Early Earth was largely of magma (molten rock) and was subject to bombardment by other massive rocky and cometary bodies. The moon is probably the result of such a collision, about 4,400Ma (i.e. 4,400 million years ago), when material resulting from the impact rebounded into space and began to orbit the Earth under the latter's gravitational attraction. In due course, as the Earth cooled, a skin – the Earth's first 'crust' – formed and small continents appeared on it which grew in size by fusing together.

It is believed that, as the Earth cooled, steam, originating from chemical reactions deep within the Earth, rose to the surface through the many volcanoes and condensed into water, forming shallow seas. Whilst the theory that water may have been brought to the Earth by impacting comets ('dirty snowballs') has not been completely discredited (cometary water in the instances when it has been analysed is isotopically different to terrestrial water), it is now considered unlikely that comets were the major provider of Earth's water.

The atmosphere of this early Precambrian Earth is thought to have been composed of a mixture of gases (nitrogen, carbon dioxide, methane, ammonia, etc.) with very little if any oxygen. It was not until soon after the beginning of the Proterozoic, about 2,300Ma, that oxygen became a significant component of the atmosphere: the Great Oxygenation Event, as it is known.

The earliest life on Earth which has so far been identified, fossilised algae in Swaziland, dates from *c.*3,500Ma. Around 600Ma, plant life in the seas developed and the first aquatic animals evolved. These were soft-bodied so they have left a very poor fossil record (see below). Plants emerged onto land *c.*420Ma, to be followed by insects and later amphibians and reptiles. Dinosaurs, descendants of reptiles and which were to dominate the world's land animals for over 100 million years, appeared *c.*230Ma. The earliest mammals date from *c.*210Ma and were to evolve rapidly in the absence of predatory dinosaurs after their mass extinction *c.*65Ma (see p.53). Birds, evolutionary descendants of reptiles, evolved *c.*170Ma.

The Earth we know today consists, in brief, of a solid inner core, mostly of iron, surrounded by a liquid outer core and it is believed that rotation of, and convection in, the liquid iron outer core generates the Earth's magnetic field. Around the core lies the rocky 'mantle' and, at about 30 kilometres below continental surfaces and 10 kilometres below the ocean floor, the 'Moho' (named after the Serbian geologist Andrija Mohorovičić) is the transition between the mantle and the crust. Some of the Earth's present heat is left over from conditions at the time of the Earth's formation – naturally, a diminishing amount – but the greater part is the result of radioactive decay within the Earth.

Continental Drift and Plate Tectonics
A casual glance at a map of the world will show a striking similarity between the shape of the coastlines of Africa and South America, across the Atlantic Ocean, as if they were part of some giant jigsaw. In 1912 the German Alfred Wegener controversially suggested that the world's continents have moved over time, and put forward the theory of what became known as 'continental drift'. Land masses have split apart and joined up from time to time as they slowly moved over the world's surface, bearing little or no resemblance to the current pattern of the continents. The most recent phase has its origin in the slow fragmentation of a super-continent, Pangaea, which began some 200 million years ago. It was not, however, until the 1960s, when the discovery of sea-floor spreading (Box 3.2) led to an explanation of the mechanism for continental drift – 'plate tectonics' – and Wegener's revolutionary theory was vindicated. (Box 3.1 provides detail of the measurement of continental drift from rock magnetism.)

Most geological events, and hence the world's geography, can be accounted for by the theory of plate tectonics. It is the unifying theory of geological science, as important to geology as is quantum theory to physics and natural selection to biology. Over many millions of years, the geography of the Earth has changed very greatly. These changes are described later in this chapter.

A detailed explanation of why and how plate tectonics function is beyond this chapter's scope (but Box 3.2 provides some greater detail), and it is sufficient to say that the Earth's surface (the lithosphere composed of the crust and upper part of the mantle) consists of a number of 'plates' of various sizes, recognised by most authorities to be 13 in number. These move or slide over another layer of the Earth, the 'asthenosphere'. There are no spaces between the plates; they move independently of one another with irresistible force so, at the plates' margins, dramatic geological phenomena such as volcanic eruptions, major earthquakes and possible resultant tsunamis can occur, as well as mountain-building. Examples of these can be found in various regions of the world. Where plates collide, crust may be destroyed; where they move apart crust must be created to fill the gap.

Shown in Figure 3.2, the Andes mountains are on the western edge of a major plate of continental crust – the South American Plate – while the adjacent bed of the Pacific Ocean is part of a plate of oceanic crust, the so-called Nazca Plate. This is moving towards South America at an average speed of several centimetres a year.

The figure shows that in the central Pacific Ocean the Nazca Plate adjoins its neighbour to the west, the Pacific Plate. These plates are moving apart and, at the boundary, magma rises from the Earth's interior and solidifies, generating new crust and contributing to sea floor spreading. This new rock contains much iron and so has a high density. The crust of

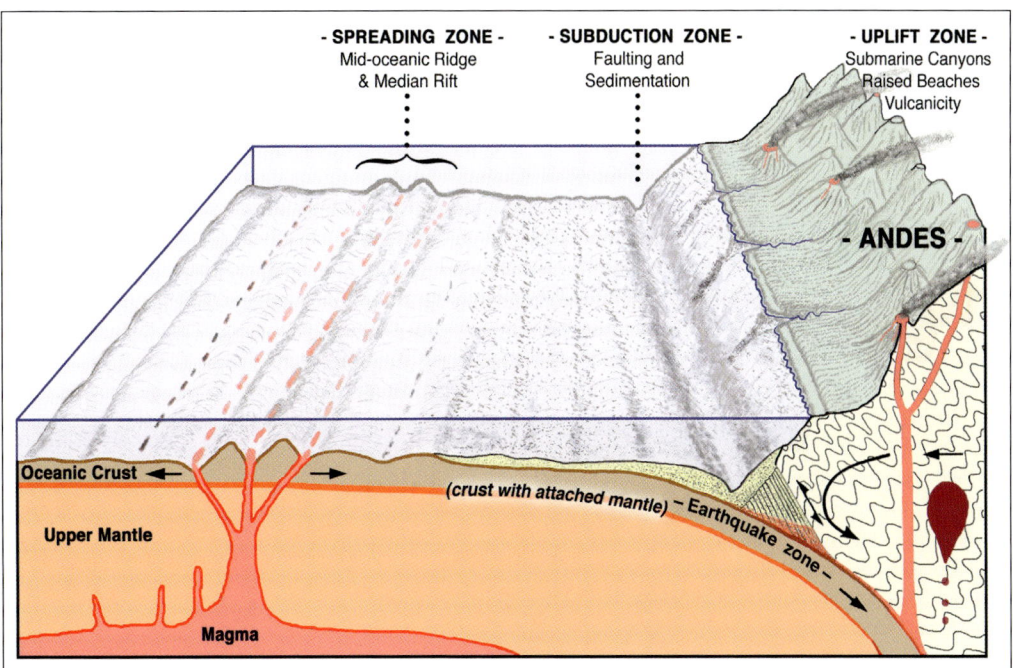

Figure 3.2 The Andes is a range of volcanic mountains resulting from the subduction of a Pacific Ocean plate. The Japanese islands are similarly formed. The diagram illustrates the spreading of the ocean floor due to the constant injection of magma at the mid-ocean ridge. (Drawn by Gerry Calderbank)

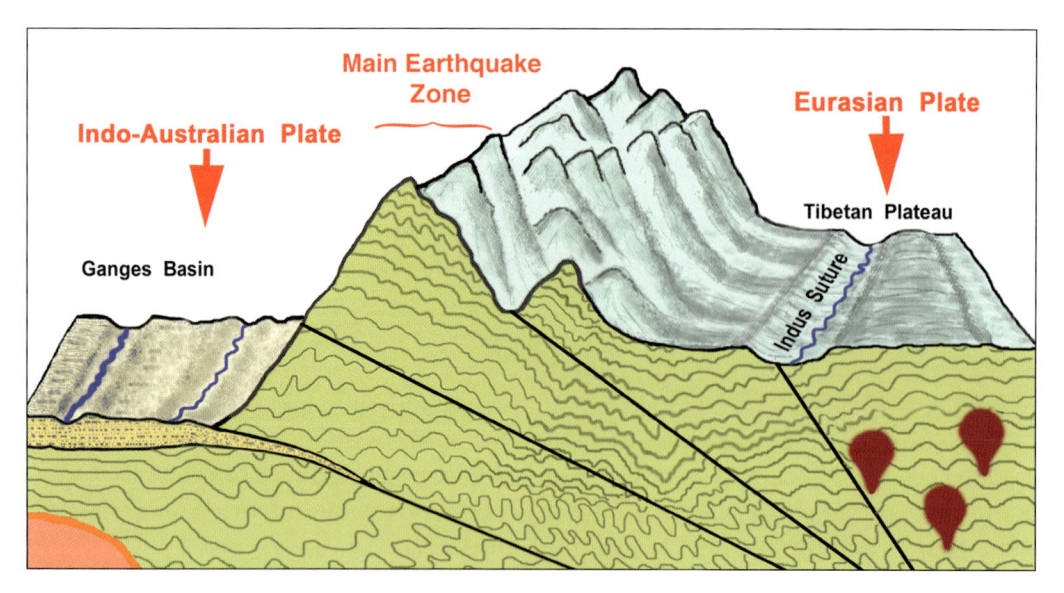

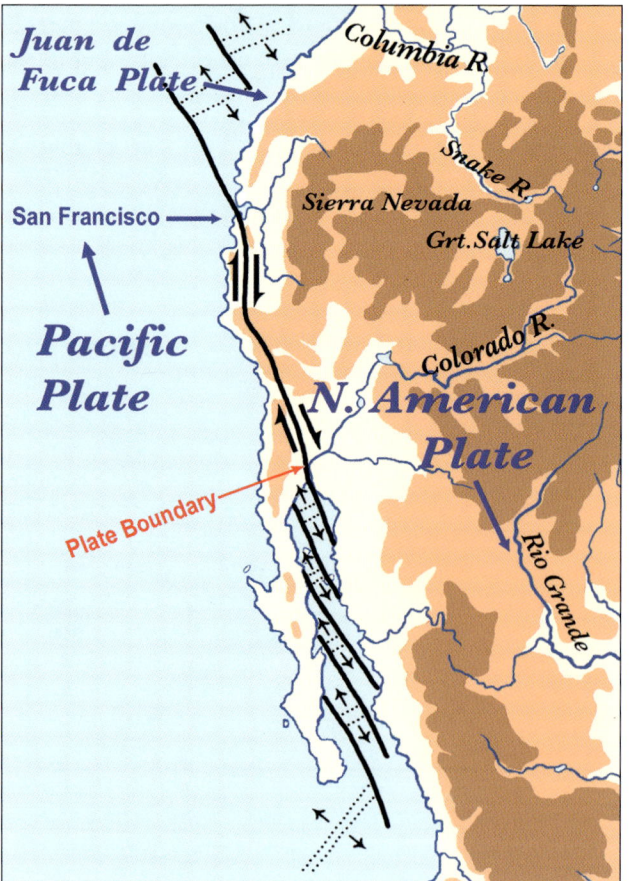

Figure 3.3 (above) The Himalayan range is the result of the collision between the Indo-Australian plate and the Eurasian plate. India started to separate from East Africa in the late Mesozoic era, moving initially at a rate of 100mm/year but slowing to 50mm/year on impact and, currently, 10mm/year. The mountains continue to rise; about 3km has been added in the last 3My.)

Figure 3.4 (left) The well-known San Andreas Fault in California occurs where three plates interact, giving rise to a fault with lateral rather than vertical displacement (a 'transform fault').
(Drawn by Gerry Calderbank)

the continents generally contains a smaller fraction of heavy elements and so is less dense. There are therefore essentially two types of crust, oceanic and continental, which differ significantly in their densities.

Where the Nazca and South American plates come into contact, the Nazca Plate, being the denser of the two is 'subducted' or forced under the other generating immense stresses deep below the Earth's surface. It is the release of the resulting strains which gives rise to the earthquakes and volcanoes to which western South America is subject. Across the Pacific, Japan is in a similar situation.

The Himalayan mountain range, which over the last 30 million years or so has been slowly growing in height, is the result of the meeting of two plates (Indian and Eurasian); but in this case they are both of continental crust. With neither plate giving way their margins buckle and are being forced upwards in a mountain-building process known as 'orogenesis', (Fig. 3.3). Terrestrial mountains are limited in height to about 10,000 metres by the inability of the Earth's crust to support the weight of higher land masses. The European Alps are also being thrust upwards under much the same conditions, while the volcanic and earthquake activity in and around the Mediterranean is further evidence of the meeting of the African and Eurasian Plates.

The San Andreas Fault in California marks the interaction of three plates, one of which is sliding past another, forming a 'transform fault' and generating stresses which result in the region's geological instability, (Fig. 3.4). Plate tectonics have also played a major part in the history of the geology of the British Isles.

Looking into the future, continuing plate movements will ensure that in, say, 150 million years the distribution of the Earth's continents and oceans will be very different to the present arrangement.

Box 3.1 **Continental Drift**

Alfred Wegener, a German geologist and meteorologist, wondered whether Africa and the Americas had once been united and had by some process moved apart during geological time. He vowed to obtain evidence from a detailed study of the geology of each side of the Atlantic to see if there were any correlations. This work and similar studies in India and Australia came together in 1912 when he proposed the previous existence of a single supercontinent called Pangaea, which had fragmented and its constituent parts had drifted away from one another. Despite the rigour of his geological evidence, the Wegener hypothesis was weakened by the lack of a convincing driving mechanism for the movement of large crustal masses.

It took until the 1950s for evidence to emerge (from studies of the Earth's magnetic field) which supported Wegener's ideas. Our planet acts as if it has at its centre a giant bar magnet. The magnetic field lines from this, shown in Fig. 3.5, have an 'inclination', the angle which the field lines make with the surface of the Earth, which varies from horizontal at the magnetic equator to vertical at the magnetic poles. Inclination, therefore, is a direct measure of latitude.

How can this phenomenon help with our drifting continents? When lava erupts from a volcano it begins to cool and crystals of iron minerals, notably magnetite (Fe_3O_4), are the first to crystallise and grow in alignment with the Earth's magnetic field lines at that point on the Earth's surface.

When the lava has fully solidified, both the direction and inclination of the local geomagnetic field have been 'frozen' into the rock. This is called remanent magnetism. A correctly orientated specimen of any lava of any age can then be analysed to yield this information and thus its latitude of formation. Any igneous rock, such as those in the Malvern Hills, can be analysed in this way and from these studies geologists have been able to determine that the Malverns Complex was formed close to 60° south of the equator, near the Antarctic Circle. The ability to find out where igneous rocks had formed provided just the evidence needed to prove that the continents had indeed moved from their original positions on the Earth's surface.

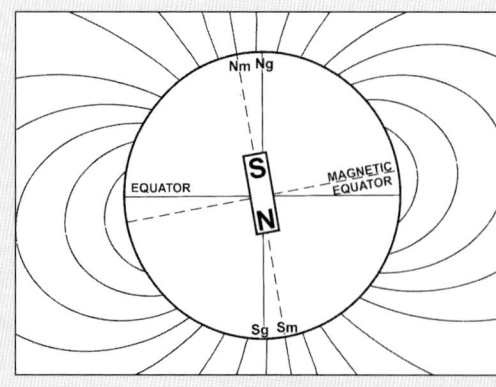

Figure 3.5 The idealised form of the Earth's magnetic field.
The Earth's magnetic field can be likened to that caused by a strong bar magnet at its core. The field lines of force are shown exiting the surface of the Earth at different angles depending on magnetic latitude, from zero at the magnetic equator to 90° at the magnetic poles, Nm and Sm (Ng and Sg are the two geographic poles). (Drawn by John Payne)

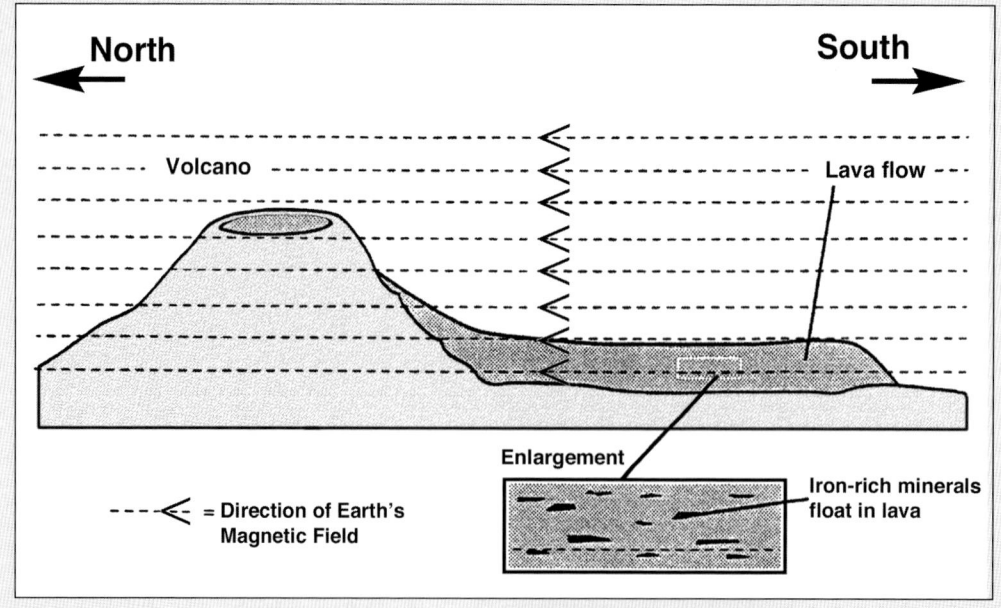

Figure 3.6 The magnetisation of iron-rich minerals.
The diagram illustrates lava erupting at the magnetic equator (where the magnetic field is parallel to the Earth's surface). This direction is preserved in the lava when it solidifies.
(Drawn by Paul Olver)

This was eventually combined with sea-floor spreading (Box 3.2) and ultimately plate tectonics to prove the mechanism for large scale continental movement.

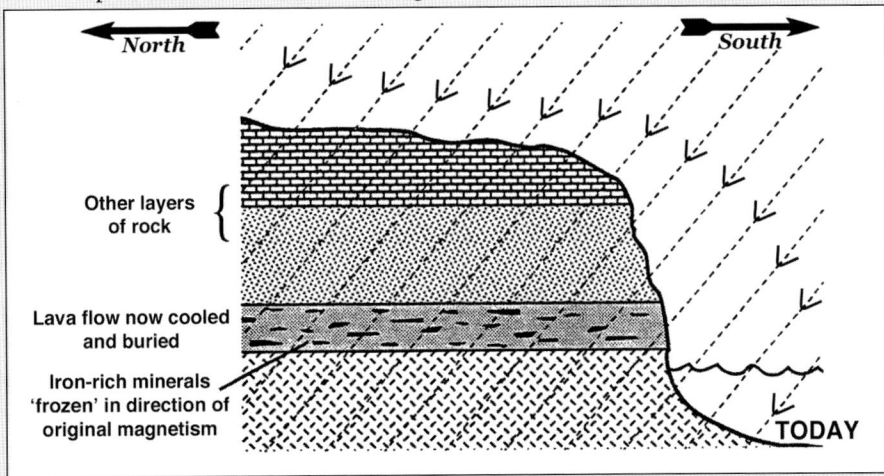

Figutre 3.7 Old lava flow now exposed in a cliff in, for example, southern Britain. The direction shown for the magnetic field is consistent with the 52ºN latitude. The lava retains its former direction of magnetisation. Once the rock is dated, this can be used to find out when and where it crystallised before continental drift brought it to its present location.
(Drawn by Paul Olver)

Box 3.2	**Plate Tectonics**

The Earth's crust is in a state of perpetual change with the lateral drift of the continents, pronounced vertical movements in mountain ranges, and the slow downward motion of crustal material in the ocean trenches.

The crust is part of the solid upper surface of our planet called the **lithosphere** which varies in thickness from a few kilometres under mid-ocean ridges to over 70km below the highest mountain ranges. In relative size, the lithosphere can be likened to the detached skin of an apple relative to the whole apple, which represents the Earth. This layer is able to move over the **asthenosphere**, part of the upper mantle, due to partial melting and plastic deformation at this level within the Earth's interior. The lithosphere is synonymous with the term 'tectonic plate' and is found to be divided into eight major plates (plus a few small ones) covering the Earth's surface. Each plate has active margins of different types but all margins share instability in terms of their frequent volcanoes and earthquakes. The movements and interactions of these plates are collectively called **plate tectonics**.

Fracturing of continents due to localised uplift and formation of rift valleys results from the development of strong convective heat flow in the asthenosphere. Rising heat flow beneath the rift leads to volcanic eruptions, melting the rocks of the Earth's mantle and initially producing **basalt**.

As the lava erupts, usually underwater in the form of pillow lavas or occasionally on land as in Iceland, the cooled mass is itself fractured by continuing heat flow motion as new lava moves up the rift fracture. Continuing this process, new ocean floor is constructed and the process of **sea floor spreading** (Fig. 3.2) is initiated. The ocean thus expands about the central rift and the continents within the plates on both sides move apart. Continents are passive in this process whereas the oceans are actively building new sea floor.

Away from the oceanic lava sources, lithospheric plates are not being created but are actively moving towards each other and being destroyed along 'destructive plate margins' (**subduction zones**). The thinner but denser oceanic plate is intermittently pushed beneath the thicker continental plate along a sloping shear zone and descends into a hotter and more compressed region. Partial melting takes place here and magma diapirs rise into the continental plate above. If these reach the surface the molten rock is erupted as 'lava'; otherwise they solidify underground as 'intrusions'. A chain of volcanic islands is formed close to the line of the subduction zone; this is an **island arc** (e.g. Japan) and is shown diagrammatically in Figure 3.8.

Active subduction eventually results in continental masses being carried towards each other. In this case, neither plate is subducted and the resultant collision leads to the deformation and upward movement of crustal rocks to form a mountain range. As an example, the collision of India with Asia 42My ago led to the formation of the Himalayas. These mountain building events are referred to as **orogenies**.

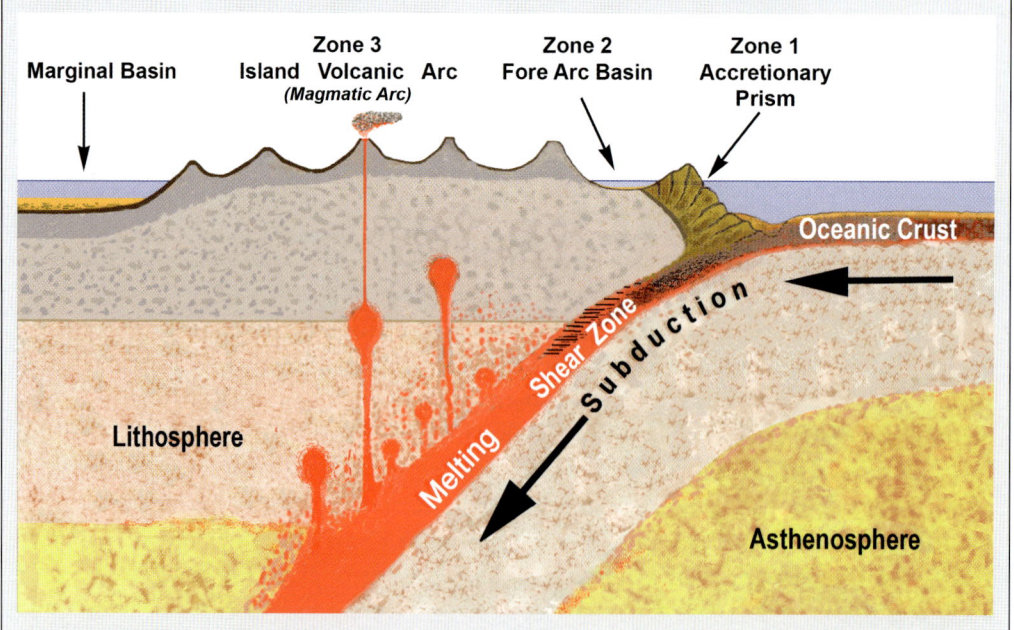

Figure 3.8 A diagram showing details of a subduction zone, with diapirs of molten rock rising to form magma chambers deep underground with subsequent volcanic activity and the generation of an arc of islands with a marginal basin (or 'back arc basin') behind. (Drawn by Gerry Calderbank)

World Palaeogeography

Plate tectonics and continental drift have profoundly affected the distribution of the continental masses. The relative placements of the continents have been traced over the past two billion years although the results from more than about one billion years ago are uncertain. The history of parts of Herefordshire extends over the last 700 million years or so, and the palaeogeography throughout most of this period is quite well understood. The local palaeogeography is described in later chapters but it is useful here to give a world and regional perspective.

The maps shown below represent a hemisphere containing what is now England, firstly with a polar view, then from 60°S and finally from an equatorial view.

During the last 500 million years, continental drift has carried the part of the crust now comprising England and Wales for a distance of roughly 10,000km, from 60°S to 50°N. The average northward speed has been around 2cm/year, about the same as the growth rate of finger nails. During this period, it has passed through a wide range of climatic zones, has been sometimes below the sea surface and sometimes above it, and has for much of the time been near the edge of its tectonic plate and so prone to involvement in plate collisions. Each of these factors may have a profound effect on the rocks and their structure and, in combination, they have led to the rich variety of British geology, including in Herefordshire.

World geography in late Precambrian times is not well known in detail, partly through the lack of fossil evidence. During this time, the earliest Herefordshire rocks, those of the Malvern Hills, were formed and this is described in the following chapter along with a more detailed description of the evolution of our area.

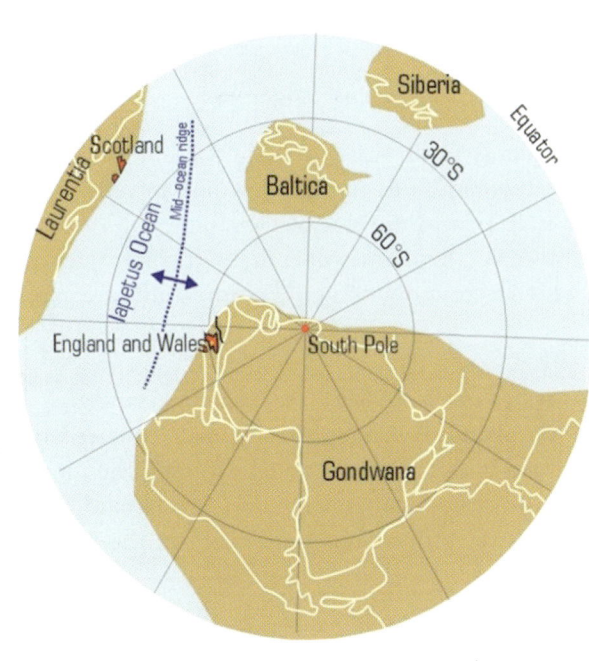

Figure 3.9 World palaeogeography in Cambrian times.
(Figures 3.9 to 3.12 drawn by Andrew and Gill Jenkinson and used with the permission of Dr Peter Toghill following their use in his books The Geology of Britain: An Introduction *(2000) and* Geology of Shropshire *2nd edition (2006).)*

The world in Cambrian times (541-485Ma) (see Chapter 5)

In the Cambrian Period, two large continents dominated the globe. Gondwana covered much of the area around the South Pole, with the proto-'England, Wales and southern Ireland' as part of a small section to be known as Avalonia (Fig. 3.9). Laurentia, the second large continent, was to the north, astride the equator, with the proto-'Scotland and northern Ireland' in it at about 20°S. Between them lay the Iapetus Ocean with a smaller continent called Baltica.

The world in Ordovician / Silurian times (485-419 Ma) (see Chapters 5 and 6)

During these periods, Avalonia, a small plate containing the present 'England and Wales', broke away from Gondwana and, starting at about 60°S, moved slowly towards Laurentia, narrowing the Iapetus Ocean while opening up another ocean to the south, now named the Rheic Ocean. In the late Ordovician Period Avalonia is believed to have collided with and joined with Baltica in what is known as the Shelveian Event (named from a Shropshire village). In the Silurian Period Avalonia moved into subtropical latitudes and the Iapetus Ocean began to close, bringing Scotland and England closer together (Fig. 3.10).

The Iapetus Ocean at this time spread from the north-west across a continental shelf on Avalonia, thereby creating a sequence of marine sediments, notably in this case the Wenlock limestone and shale beds, which today can be seen in Herefordshire to the west of the Malvern Hills, in the Vale of Wigmore and in the Woolhope area to the east of Hereford city. They are also prominent on Wenlock Edge in Shropshire. These limestones are particularly rich in fossils of marine fauna, including corals, which were common in the warm, shallow tropical waters at this 30°S latitude.

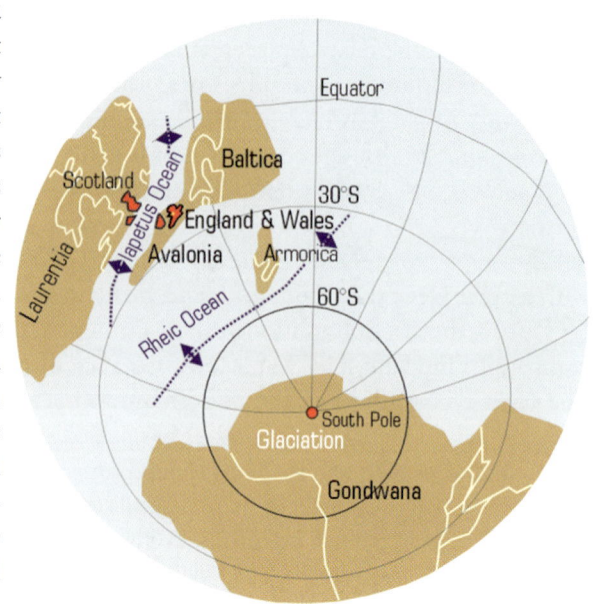

Figure 3.10 World palaeogeography in Silurian times

The world in Devonian / Carboniferous times (419-299Ma) (see Chapters 7 and 8)

During the later stages of the Silurian Period, Avalonia, having already made contact with the plate known as Baltica, eventually collided with Laurentia, thereby joining the proto-'England, Wales and southern Ireland' with the proto-'Scotland and northern Ireland' (Fig. 3.11). As a result, huge mountains of Himalayan proportions were raised in the process known as the Caledonian Orogeny.

With the collision of the plates, the formation of what is known as 'The Old Red Sandstone Continent', and the creation of the Caledonian mountain system, the inevitable process of erosion set in. Over millions of years, in semi-desert conditions and with flash flooding, vast amounts of sediment were washed from the mountains, eventually forming great thicknesses of sedimentary rocks. The Old Red Sandstone was laid down at this time as huge rivers poured southward from the mountains, depositing their sediments either in or near to the Rheic Ocean. Today, the north-east/south-west trending mountains of Wales, the Lake District and Scotland are the remnants of these once much greater mountains of Devonian times (since then eroded, buried and now re-exposed).

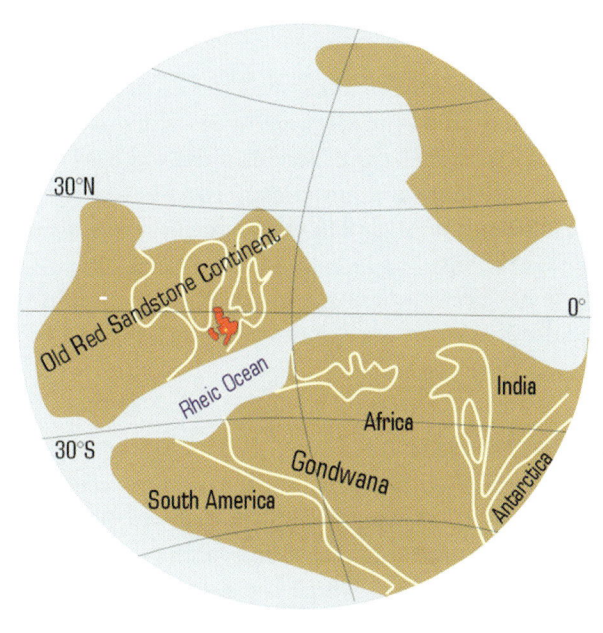

Figure 3.11 World palaeogeography in Devonian / Carboniferous times

The area of Herefordshire was part of an extensive coastal plain which received vast amounts of these sediments from the mountains. Today these form the county's signature 'Old Red Sandstone' sedimentary rocks, conspicuous in older buildings throughout the county and in the red soils so characteristic of much of the county's agricultural land.

In due course, during the Carboniferous Period the combined plates moved further north into tropical latitudes, our area crossing the equator at about 320Ma. A sequence of sediments was deposited as Carboniferous Limestone, Millstone Grit and Coal Measures (but see Chapter 8 for current terminology), the last being derived from the lush vegetation which was then widespread. At this time 'Herefordshire' was to the south of an upland area named by geologists as St George's Land and today its Carboniferous deposits are in evidence in the south of the county. Other local and associated Carboniferous outcrops are to be seen at Titterstone Clee in southern Shropshire, the Forest of Dean in Gloucestershire and the South Wales Coalfield.

The world in Permian / Triassic times (299-201Ma) (see Chapter 8)
During this period the Rheic Ocean closed as the Devonian/Carboniferous 'Old Red Sandstone Continent' recombined with Gondwana to form the super-continent which geologists have named Pangaea. This collision was accompanied by the Variscan Orogeny which, in our area, affected southern Britain and central Europe. The now-combined 'Britain' meanwhile became an inland part of the continent, north of the equator and in desert conditions for a second time. Deposition in these deserts

Figure 3.12 World palaeogeography in Permian / Triassic times

gave rise to the New Red Sandstone formations so conspicuous in Worcestershire and the Midlands.

In the following Jurassic and Cretaceous times a major event was the opening of the Atlantic Ocean, separating the American plates from Europe and Africa along the line shown in Figure 3.12. For most of this time the area including Herefordshire was below sea level in a shallow sea, with the deposition of sandstones and limestones. These have all been eroded away from Herefordshire in the last 60 million years but they remain in south-east England including the Cotswolds and the Chalk hills.

For the last 50 million years or so there has been little northward progress of our area and, because of the tectonic emergence and widening of the Atlantic, a drift to the east has taken place and is continuing.

Minerals, Rock Types and the Rock Cycle

Virtually all rocks consist of more than one mineral. A mineral is a naturally occurring substance whose atomic structure gives it a particular crystalline form and which has an individual chemical composition or range of compositions. For example, the commonest mineral in the Earth's crust is quartz (SiO_2; silicon dioxide) which is made up of one atom of the element silicon for every two atoms of the element oxygen.

For geological purposes, rocks can be divided into three major categories: sedimentary, igneous or metamorphic, depending on their manner of formation.

Sedimentary rocks

These are the most commonly occurring rocks at the surface and consist of sediments that were laid down on land or under water. These were later cemented into identifiably separate 'beds' under the pressure of successive later-deposited layers on top of them. The boundaries of the beds are shown by some change in the nature of the rock because of an alteration in the conditions of deposition. The sediments may be of organic origin, for example forming coal from decomposed vegetation, or limestone and chalk. Chalk is essentially a pure form of limestone and is typically laid down at the rate of 100 metres in five million years from organic calcium carbonate found in the remains of marine organisms. Sediments may also be of inorganic origin such as eroded fragments ('clasts') of existing rocks. The rocks thus formed are classified by the average clast size as mudstone, siltstone, sandstone or conglomerate. Sedimentary rocks can in practice consist of a number of materials of organic and inorganic origin.

Igneous rocks

These are formed from solidified magma. Magma is molten rock held under pressure and heat in the Earth's crust and upper mantle. Structural flaws in the crust may allow magma to be forced to the surface where, in the case of volcanoes, it emerges as lava and is described as 'extrusive' rock. Should the magma solidify when buried within the crust, it is termed 'intrusive'. An example of the former is the common igneous rock basalt which, erupting on to the Earth's cool surface, has solidified relatively quickly – over, say, two weeks. Therefore, its crystals have not had time to grow; they are usually invisible to the naked eye. Granite, on the other hand, is an intrusive rock which has solidified comparatively slowly in the crust – perhaps taking a million years to do so – and which later erosion has often exposed on the Earth's surface. Its crystals have therefore had time to grow and are clearly visible.

Igneous rocks may form 'sill' and 'dyke' structures just below volcanic centres. A dyke is a more or less vertical intrusion of igneous material which has been forced through fissures in a pre-existing rock. A sill is an igneous intrusion emplaced between beds of rock in usually a more nearly horizontal plane (Fig. 3.18).

There are hundreds of different types of igneous rocks. They are classified as being 'basic', 'intermediate' or 'acidic' depending on how much silica they contain. Silica makes up some 45% of basalts, which are basic rocks, but about 70% of acid rocks such as granites. (The misleading but now conventional use of the terms 'acidic' and 'basic' stems from early misconceptions about the origin of such rocks. Approximate synonyms are 'felsic' and 'mafic' respectively.)

Volcaniclastic rocks are formed by the accumulation of fragments and particles from volcanic eruptions. An example is bentonite, a clay derived from fine volcanic ash.

Metamorphic rocks

These are rocks of either sedimentary or igneous origin whose crystallisation has been changed through pressure from tectonic forces and/or heat from igneous sources. When plates collide, tectonic forces can result in 'regional metamorphism' over a considerable area, whilst the heat from molten igneous rock can cause local 'contact metamorphism' in the surrounding rocks. When the changes result from the heat and pressure caused by local tectonic movements such as faulting, the process is known as 'dynamic metamorphism' and its effects are usually localised. (This is the chief metamorphic class in the south Malverns.)

Most of the oldest of the Earth's rocks have been metamorphosed more than once. When this process can be detected over a considerable area it is often a sign of a mountain-building episode in the remote past whose rocks have since been worn away to a greater or lesser extent. Limestone is metamorphosed into marble, and mudstone into slate which, if subjected to further pressure, changes into schist and finally, if nearly melted, into gneiss. Metamorphosed sandstone forms quartzite, one of the hardest and most durable of rocks.

Igneous rocks are discussed further in Box 4.1 and metamorphic rocks in Box 3.3. Photographs of some local igneous and metamorphic rocks are shown in Chapter 4.

Box 3.3	Metamorphism and Metamorphic Rocks

All metamorphic rocks were once igneous or sedimentary rocks. They have been transformed by temperature and pressure changes within the Earth's crust. These changes involve the re-crystallisation of existing minerals (e.g. a marble) or the appearance of new minerals stable under the new temperature/pressure conditions (e.g. kyanite schist). Deformation processes associated with metamorphism and the growth of new minerals produce new textures within the rock mass (e.g. schistosity, a structure of layered mica crystals in schists). Former bedding planes and fossils, both of which are characteristic of sedimentary rocks, are gradually destroyed by increasing metamorphism. Metamorphic processes take place at temperatures below the melting point of the rock mass and involve crystallisation in the solid state without the addition or loss of any new material except for volatiles (H_2O and CO_2). Metamorphism usually requires the geothermal gradient to be above the normal for our planet (20°C per km) or, in the special case along subduction zones, above 10°C per km.

There are three main categories of metamorphism based on their field occurrence.

Regional metamorphic rocks outcrop over large areas and are produced at depth by rising heat and pressure within major mountain ('orogenic') belts. The pressure element and the growth of new minerals produces new textures or fabrics such as **cleavage** in slates and phyllites, **schistosity** in schists and **foliation** in gneisses as the temperatures and depths of formation rise.

Contact metamorphic rocks occur in localised zones adjacent to igneous intrusions and are primarily the product of rising temperatures. The highest grades of contact metamorphism are therefore found nearest the igneous contact. The massive, often flinty, rocks found in these zones are called **hornfels**.

Strong deformation processes caused by increasing shear pressures generated by colliding plates produce both faults and thrust planes. At shallow levels such structures are dominated by fragmentation of the rocks involved and the production of fault breccia. However, at greater depths, under both high load pressure and increasing shear stress, the rocks undergo local re-crystallisation to produce a *cataclastic* metamorphic rock. This is **dynamic** metamorphism.

In dynamic metamorphism, if the surrounding rock is a mudstone, shale or schist containing a high proportion of clay minerals or similarly structured micas, these minerals re-crystallise parallel to the fracture surface. The result is **phyllonite**, similar to a regionally metamorphosed phyllite/schist, but which splits less readily along the mica layers.

Massive, mechanically strong rocks such as granites and quartzites give rise to very fine-grained, flinty textured **mylonites**. Fragments of the fractured rocks are often found smeared out along such zones and, if they consist of one crystal, are called porphyroclasts.

Extreme cataclastic metamorphism leads to fine-grained mylonite devoid of any included rock slivers and, with increasing frictional heat, local melting on the fracture zone is initiated. On solidification, this glassy material is called **pseudotachylyte**.

The Rock Cycle

Fig. 3.13 illustrates the basic features of what is known as the Rock Cycle – the continuous natural processes by which rocks are broken down and recycled to appear in new forms.

All rocks, even the hardest, will in time crumble through exposure to the weather and physical and chemical processes. The resulting fragments (clasts) may be transported elsewhere by gravity, water, ice in the form of glaciers, or wind to be deposited and buried as sedimentary rocks. These, as we have seen, may be metamorphosed by pressure and/or heat and may melt to form molten igneous rocks. By the erosion of overlying rocks, these may later be exposed at the Earth's surface where they are again broken down into fragments and the rock cycle is complete. The cycle can, however, be short-circuited at any stage by 'uplift' and deformation of the Earth's crust which may be regional in character as the result of tectonic forces, or may be more of a local origin. When this occurs, rocks of any type can be forced to the surface.

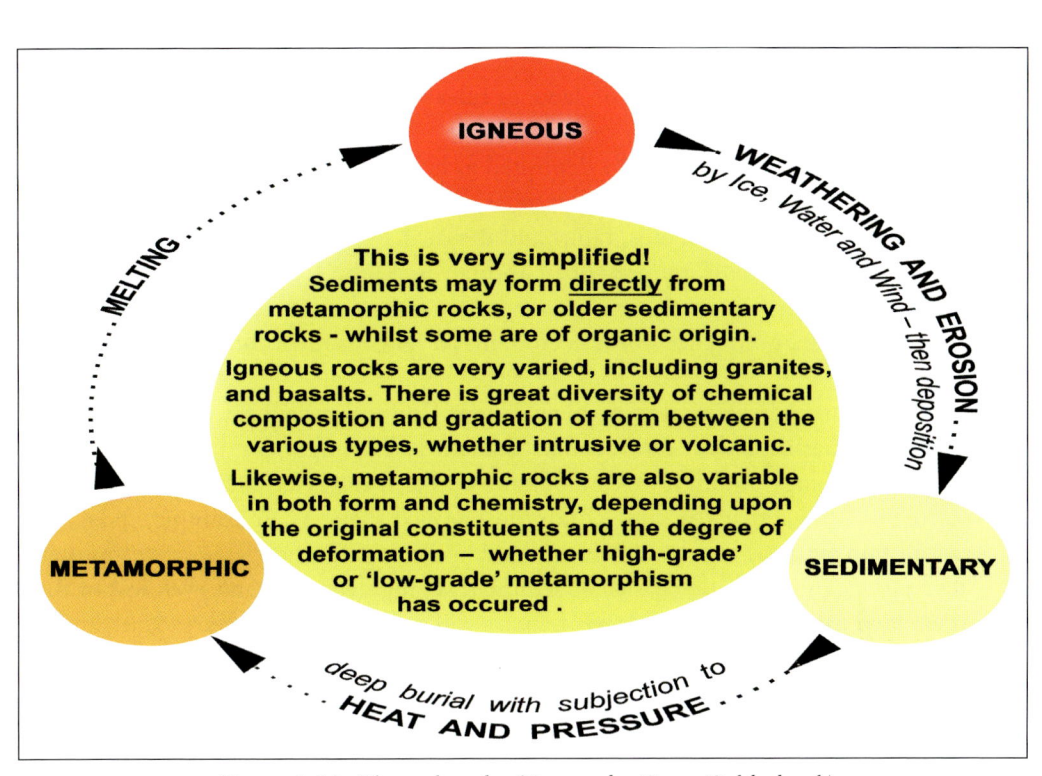

Figure 3.13 The rock cycle. (Drawn by Gerry Calderbank)

Geologising: reading the rocks

It is possible from detailed examination of rocks in the field, under the microscope and by a variety of laboratory techniques, to gather information about the remote past. The climatic and geological environment at the time a particular rock was formed can be deduced, and any fossils within it may provide evidence for evolutionary studies. Geologists may use scientific techniques to help them unravel the past but, in the field, careful observation alone will enable them to tease out a considerable amount of information about a particular rock and its place in geological and environmental history.

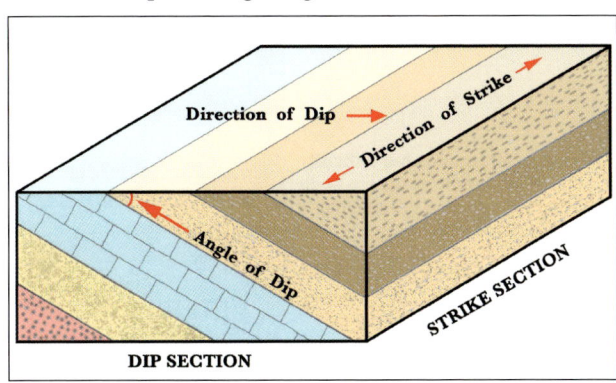

Figure 3.14 Dip and strike. (Drawn by Gerry Calderbank)

The field geologist uses a 'compass clinometer' to detect some of the basic characteristics of beds or strata. Recording the 'dip' establishes the steepness of the angle of the bed (or stratum) by measuring the relative angle of a bedding plane (a plane which separates individual beds or strata) to the horizontal. The compass clinometer will also record the direction of the 'strike' of the rock unit as illustrated in Figure 3.14.

Stratigraphy

Geologists have developed a number of 'laws' or rules from the study of the strata of mostly sedimentary rocks, for example:

(a) Sedimentary rocks are generally deposited as horizontal layers.

(b) In a sequence of rocks, the youngest rocks will normally be at the top and the oldest at the bottom (but see below, 'Folds, faults and thrusts').

(c) Where rock fragments are present in another rock, the fragments will be the older of the two.

(d) Where a rock or sequence of rocks contains an igneous intrusion, the intrusion will be the younger.

(e) A further major geological principle is that of Uniformitarianism, which introduced this chapter and applies much more widely than to just stratigraphy. It states that if we understand the geology of today, we can look for similar features left behind in the geological record and deduce that the same conditions applied then. As an example, today land areas like Britain are areas which are being eroded by rivers and other destructive forces and on whose surface there is little chance of sediments being deposited and even less of their being preserved for posterity. This state of affairs will continue until the area starts to sink and there is a possibility of any new sedimentary deposits being covered by later material and so preserved as layers of sedimentary rock. The period of emergence and erosion as a land area will be marked in the geological record as a gap in the sequence of deposits (an *unconformity*).

In general, an 'unconformity' (Fig. 3.15) exists when there is a significant break in the stratigraphy, such as when a Devonian rock has been deposited directly on top of an Ordovician one; a rock of the intervening geological period, the Silurian, may have been

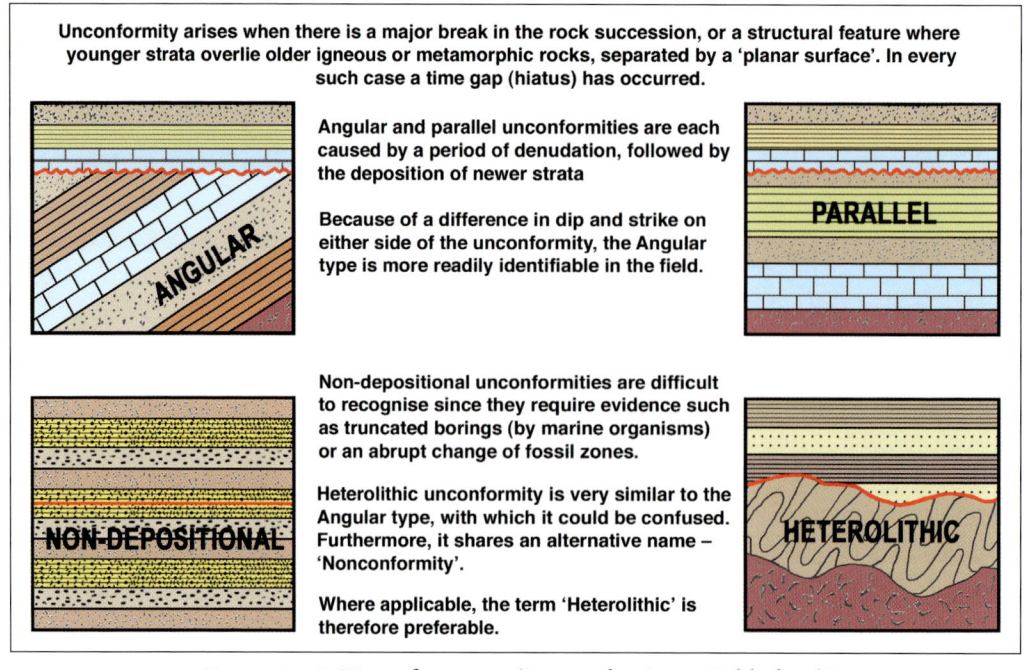

Figure 3.15 Unconformities. (Drawn by Gerry Calderbank)

| Box 3.4 | **Cross-bedding** |

When sand is deposited from currents in shallow water, shoals and sandbanks are often built up. The front of deposition advances in the direction of the current. The bedding of a growing sandbank follows the gently sloping surfaces on which the sand is dropped.

Figure 3.16 Cross-bedding. (Drawn and photographed by John Payne)

completely eroded or perhaps was never deposited. Unconformities may represent breaks in deposition of very large or very small duration.

'Cross-bedding' is an instance of a very minor unconformity, usually in sandstone. Here, the movement of sediment grains into dunes by wind or water can cause successive bedding planes to be deposited at different angles to each another – typically, beds laid down horizontally above and below a cross-bed lying at an angle to the horizontal. From the orientation of the cross-bedding, the direction of the wind or water flow (the palaeo-current) driving the process can be found. This is particularly useful in analysing sediment deposited by rivers.

'Folds' (Fig. 3.17) occur when rock strata in the Earth's crust are *bent* under high heat and pressure and can be spectacularly complex in shape. An 'anticline' is a convex-upward fold in the rocks; a 'syncline' is a concave-upward fold.

'Faults' (Fig. 3.18) of a number of different types are the result of forces acting on rocks which *fracture* with movement on either side of a vertical or inclined fracture. If the fracture is the result of extensional forces, the rock above the fault plane can slip downwards under

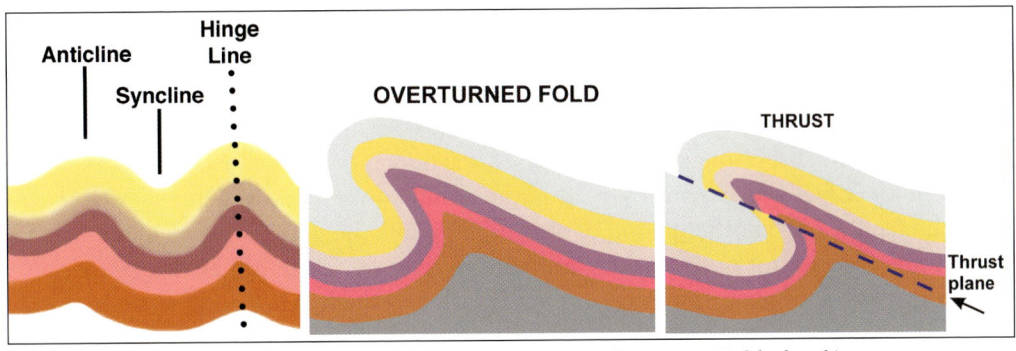

Figure 3.17 Folds and thrusts. (Drawn by Gerry Calderbank)

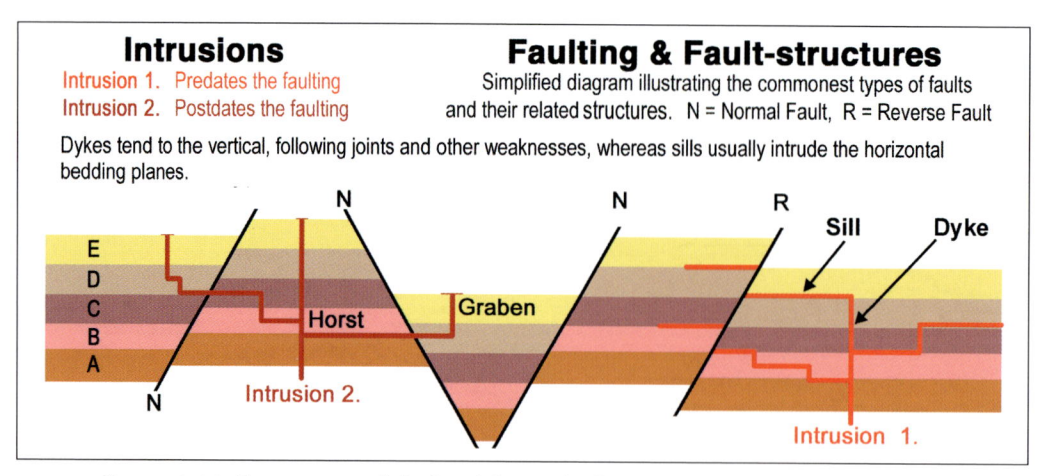

Intrusions
Intrusion 1. Predates the faulting
Intrusion 2. Postdates the faulting

Faulting & Fault-structures
Simplified diagram illustrating the commonest types of faults and their related structures. N = Normal Fault, R = Reverse Fault

Dykes tend to the vertical, following joints and other weaknesses, whereas sills usually intrude the horizontal bedding planes.

Figure 3.18 Intrusions and faults, dykes and sills. (Drawn by Gerry Calderbank)

gravity. The result is known as a 'normal fault'. When the forces are compressive, rocks are pushed *up* the plane of the inclined fracture, so overriding other rocks; this can lead to older rocks lying above younger ones. When the gradient of the fault plane is small, this is called a 'thrust'; otherwise it is known as a 'reverse fault'. If parts of an intrusion lie on both sides of a fault, Figure 3.18 indicates how it may be judged which is the older.

'Strike-slip' faulting occurs when the movement is in the direction of the strike (Fig.3.14) of the fault; that is, the movement is approximately horizontal.

Once a fault has formed, the broken rock presents a long-lasting zone of weakness. Any subsequent tectonic forces acting on it, perhaps hundreds of millions of years after its formation, are likely to cause further movement on the fault rather than in the neighbouring rocks. Such 'reactivated' faults have been significant in Herefordshire's geological story.

Some effects on the landscape of variations in erosion resistance and of faulting are shown in Fig. 6.12, in the chapter where they have greatest relevance.

Ice Ages

Ice ages are defined as extended lengths of time during which large parts of the Earth are covered by sheets of ice. These periods of severe cold, 'glacial periods (glacials)', are punctuated by stages of relative warmth, 'interglacial periods (interglacials)'.

The Earth has been subjected through its history to a number of ice ages, including an episode known as 'Snowball Earth', about 650 million years ago, when it seems that the whole planet was for many millions of years subject to an extreme ice age. Subsequent ice ages have affected the higher latitudes of the Earth and it is believed that these latitudes, including Great Britain, could now be in an interglacial of the latest ice age. This is the most recent phase, beginning about 10,000 years ago, of a cycle of climatic changes which have been identified as occurring over the last two million years and which may well in due course be replaced by another period of glaciation. While the possible *causes* of ice ages are beyond the scope of this chapter, some of their *effects* on the Earth can be described.

During periods of glaciation, the sea level falls due to the locking up of water in the ice sheets. In the glaciated areas the weight of the great thickness of ice, perhaps as much as four kilometres, has the local effect of depressing the Earth's crust tending to reduce the apparent fall in sea levels relative to the land. As the ice melts during an interglacial, sea level rises and, much more slowly, the depressed crust rebounds – 'isostatic adjustment' – because it is relieved of the weight of the ice. This reduces somewhat the local rise in sea level. Such changes in sea level, resulting from a late-Ordovician glaciation, had major significance for the deposition of the Silurian rocks of Herefordshire (Chapters 5 and 6).

When a glacier melts at its lower end – the 'snout' – either in the higher temperatures of a lower altitude or as it retreats in an interglacial, it deposits 'till' (also known as 'boulder clay' or 'drift') which it has ground from the valley floor. This constitutes a 'moraine', now identifiable as a low rounded hillock or hillocks, and a succession of these can indicate stages in the retreat of a glacier. Also, as a glacier melts, it deposits rocks, which it has been carrying, perhaps over many kilometres, on the terrain over which it once flowed. These are 'erratics' – detached rocks now found lying on a different bed rock. 'Kettle holes', often identifiable as ponds or depressions in the ground, are formed when a mass of ice is buried in a moraine so that when it melts the ground above it caves in.

Recent ice ages have caused a number of rivers to alter their course dramatically. A glacier may, for instance, form a dam across a river course and create a lake. With a rise in water level of the lake the water, unable to flow along its former course may force a new way as a river through the landscape.

Well-known glacial landscape features such as U-shaped and hanging valleys, seen in more heavily glaciated areas, are absent in Herefordshire.

Palaeontology: the study of fossils

Fossils are the remains of long-dead animals and plants, or traces such as burrows left by them. It is believed that only about one per cent of organisms that have ever lived on Earth have left fossils as proof that they once existed and these must be at least 10,000 years old to be described as fossils.

In Precambrian rocks, fossils are few and far between but in the mid-Cambrian the evidence from fossils indicates that life on Earth had suddenly become widespread and varied. This phenomenon is known as the 'Cambrian Explosion', but most geologists now deny that it ever took place, their point being that earlier soft-bodied organisms left very few fossils compared to the number left by evolving hard-shelled organisms in the Cambrian.

Fossils provide data for the relative dating of rocks and of the evolution of life on Earth – inadequate as the fossil record is – as well as of past climatic conditions and environment. Graptolites, for example, thrived in deep-sea conditions, trilobites in shallow water. Corals need clear water with a temperature of at least 20°C as well as sunlight, i.e. warm, shallow water. Interestingly, present day corals evince 365 daily growth rings per year while the 400-million-year-old fossils of mid-Devonian corals reveal about 400 rings per year. It has thus been deduced that the Devonian year was 400 days long and the Earth must have been spinning faster.

It is largely from palaeontology that it has been established that there have been several mass extinctions of life on Earth. The best-known ones are that at the end of the Permian (261Ma), leading to the loss over a long period of perhaps 95 percent of species, and the one at the end of the Cretaceous (65Ma) when about 80 percent of species disappeared. This catastrophic loss included the dinosaurs which were probably destroyed by a combination of events, including the consequences for the Earth's environment of a massive meteorite impact. Mass extinctions can derail the natural process of evolution on Earth; dinosaurs had been around for some 150 million years before their extinction only 65 million years ago. Without that calamity they might still be dominating the land animals of today, including the now predominant mammals.

Fossils are most commonly found in sedimentary rocks laid down in 'quiet' conditions where the hard parts have a chance of remaining unbroken, such as limestones on the floor of shallow seas. Conglomerates (water-formed, rounded rocks or pebbles cemented by another material) are unlikely to have allowed fossilisation to take place. Metamorphism will probably have destroyed or distorted any fossils present in the parent rock. Igneous rocks, with their origin deep within the Earth, do not contain fossils. Thus, the marine Silurian rocks are Herefordshire's most prolific sources of fossils.

The process of fossilisation can take several forms resulting in different types of fossil. In the case of body fossils the basic process is the replacement by minerals of the hard parts of an organism such as its skeleton, shell or a plant stem as it decays. After death and the rapid decomposition of the soft tissues, the remaining hard parts are buried in sediments and are very slowly replaced by minerals dissolved in the ambient water. Soft-bodied organisms can occasionally be preserved as fossils when, amongst other factors, conditions allow more or less immediate burial in sediment after death. Sites yielding exceptional fossils, such as abundant soft-bodied forms, are recognised as *Lagerstätten*; these are very rare but one has been discovered in Herefordshire (see Chapter 6). In the case of trace fossils, minerals fill a void such as a burrow, or footprint. Amber, the transparent, hardened resin or sap from coniferous trees, can contain beautifully detailed fossils of insects which have been trapped within it.

Fossil fuels

Oil and natural gas are products of the decay of plants and organisms buried in sediments usually found on ancient sea floors. Heat, pressure and the chemical environment transform such remains into hydrocarbons, a process which needs a period of at least five million years. In this way, the original dead matter is transformed, depending principally on time and temperature, into either oil of various constituents and viscosities, or a variety of gases.

Coal, a carbon-rich sedimentary rock, is the product of decayed land plants which have been buried in an oxygen-impoverished environment, typically swampy conditions subject to changes of water level. This environment existed in parts of Britain, but not Herefordshire, during the late Carboniferous. There is thus no coal in the county, except for small scale working at Howle Hill (see Chapter 8), but this has not inhibited occasional ill-founded explorations for it (see Chapter 10).

Further reading
Fortey, Richard *The Earth; An Intimate History*, HarperCollins (2004)
Park, G. (2nd ed.) *Introducing Geology: A Guide to the World of Rocks*, (2010)
Rothery, David A. *Teach Yourself Geology*, Hodder & Stoughton (1997)
Toghill, Peter *The Geology of Britain: An Introduction*, Swan Hill Press (2000)
Whittow, John *The Penguin Dictionary of Physical Geography*, Penguin Reference Books (1998)
Dictionary of Geology & Mineralogy, McGraw-Hill (2003)

4 THE PRECAMBRIAN ERA
– HEREFORDSHIRE'S FOUNDATION

The Malvern Hills : A geological journey

The Malvern Hills comprise a ridge 12km in length and 425m in maximum height forming an impressive eastern edge to the county of Herefordshire. The hills are never more than 1km in width and separate the flat lying Permo-Triassic plain of the Midlands to the east from the gently folded Palaeozoic (mainly Silurian with some Cambrian) country to the immediate west (Fig. 4.1).

Figure 4.1 (below) A view of the Malvern ridge from the south. To the right (east) is the flat Severn Valley, with its deep underlying rocks downthrown 2500m along the East Malvern Fault Line, which runs immediately to the east of the hills. The nearest hill in view is Raggedstone Hill with the large Hollybush quarry just beyond it. To the left (west), the near ground is the northern part of the valley of Whiteleaved Oak, an area of Cambrian and Ordovician rocks bounded on the north by Obelisk Hill, on which the white pillar of the obelisk is visible.
(© Derek Foxton Archive)

The area of the Hills is only partly in Herefordshire. Worcestershire and Gloucestershire also have their shares. Nevertheless, this chapter deals with the Hills as a whole. The geology does not change at the county boundaries and some geological features are best described from sites on the Hills but outside Herefordshire, although close to the county boundary.

The Malvern Hills are a quintessential part of a typical English landscape. In geological terms, however, they are very much strangers in that landscape. The rocks seen today are the product of extensive activity close to the Antarctic Circle about 700 million years ago involving the collision of a small tectonic plate with the large southern continent of Gondwana (Box 3.2 and Fig. 4.3). The igneous rocks formed then, deep underground, make up today's Malvern Hills. The variety of igneous rock types, of Precambrian age, which forms the Malvern Hills is collectively known as the Malverns Complex. A good selection of these widely differing rocks frequently appears in the walls of local buildings.

A stop at St James Church in West Malvern (Fig. 4.2; SO 763 461) overlooking the gently rolling landscape of Herefordshire reveals a typical local use of the stone. The church shows the characteristic use of angular blocks of variegated igneous rocks closely fitted together and cemented by prominent mortar layers. The local Precambrian stone was not easily worked but was in plentiful supply at the time of building. In fact, quarrying was one of the main occupations for Malvern residents, reaching its peak in late Victorian times and lasting until 1977, when the final quarry closed at the Gullet in the south of the hills.

Close to the church on its south side is a natural spring enclosed in decorative stonework. Malvern springs have been renowned for medicinal purposes since at least 1622. Analysis of selected Malvern waters by Dr John Wall in 1756 drew attention to their purity and their effectiveness in water cure therapies ('hydropathy'). The combination of pure waters, pure air and the easily climbed Malvern ridge eventually proved irresistible to the newly mobile Victorians who flocked to the area from the 1840s onwards. The whole area experienced its greatest period of growth, accelerated by the coming of the railway in 1860.

Leonard Horner, the father-in-law of Sir Charles Lyell, produced the first major contribution to Malverns geology[1] and the intervening 200 years has seen the area not only become one of the classic areas of British geology but also attract considerable attention from archaeologists with its prominent Iron Age hillfort surmounting the Herefordshire

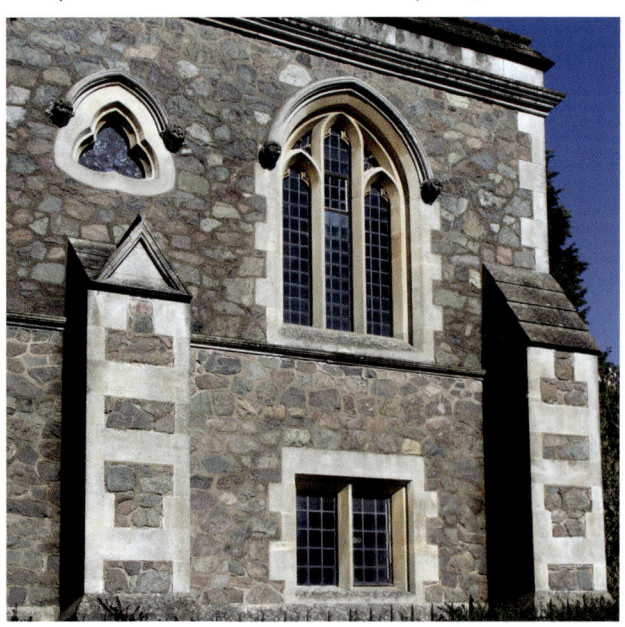

Figure 4.2 St James Church, West Malvern, showing the characteristic use of Malverns Complex igneous rocks as its main building stone. The quoins are of Cotswold oolitic limestone. (© John Stocks)

Beacon and local finds of characteristic Malvern pottery. Down through the centuries, the hills have inspired artists, writers, poets and composers, not the least of whom was Sir Edward Elgar who regularly walked, cycled and flew kites in and around its rocky ridges.

This chapter continues with an account of the origin and development of the Malvern Hills over a long period of geological time. This is followed by a description of the rocks which make up the hills, as seen in a north to south traverse of the range. The chapter ends with a mention of the three small other exposures of Precambrian rocks in, or close to, Herefordshire.

Evolution of the Malvern Hills

The development of the present Malvern Hills is the result of events in several geological periods, events described in this section. The more general developments in Herefordshire within these geological periods are described in later chapters. Further description of the origins of the Malvern Hills is given in Chapter 8.

The Precambrian Period – before 542Ma

During the late Precambrian, at around 700Ma, a collision took place between a small tectonic plate, carried southward with the oceanic crust of the Iapetus Ocean, and the northern edge of the Gondwana continent which was located near the south pole. This resulted in the 'Cadomian' orogeny. Subsequent subduction of the oceanic crust at the continental edge between 700 and 570Ma gave rise to a chain of volcanic mountains, the Cadomian island arc (Fig. 4.3). This was a string of volcanic islands like Japan or the Caribbean today (Box 3.2). The island arc was made up of a series of volcanoes fed from large magma reservoirs in the upper crust together with sedimentary rocks such as sandstones and shales deposited between the volcanic islands, some of which were dragged down with the subducting plate. Cooled and solidified magma within the magma chambers formed igneous crystalline rocks. These igneous rocks and some of the sedimentary rocks lay deep underground in the hot and highly stressed subduction zone and some were

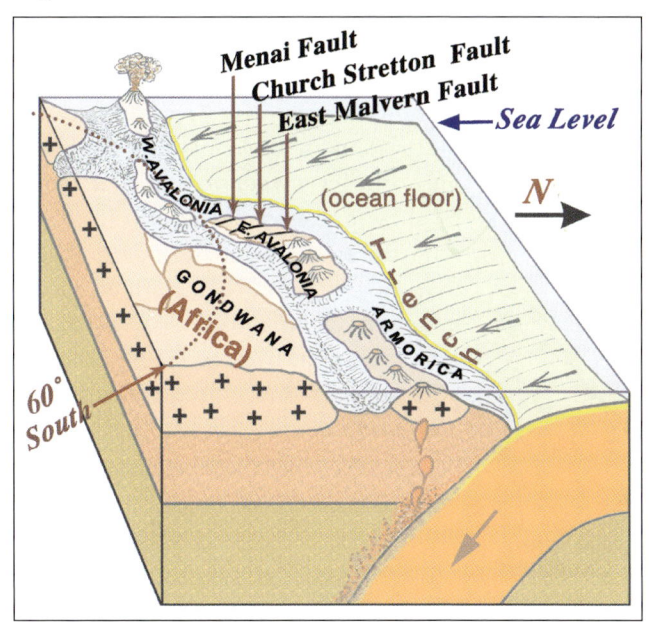

Figure 4.3 The movement and subduction of the oceanic plate (on the right side of the diagram) drove the small plates of Avalonia and Armorica against Gondwana. The subduction led to the generation of volcanoes in the small plates, thus forming the Cadomian island arc. (Drawn by Gerry Calderbank from a conceptual sketch by Dave Green)

converted to metamorphic rocks. The igneous and metamorphic rocks from a small part of the island arc have, over a long time, been brought to the surface and today form the Malvern Hills. This group of rocks is collectively known as the Malverns Complex and this part of the island arc is known as Avalonia although it became a separately identifiable unit only in Ordovician times.

The metamorphosed sediments and remnants of these most ancient rocks of the Malvern Hills occur south of the Wyche and increase in proportion towards the southern Malverns. The volcanoes have long since disappeared in the long period of erosion at the end of the Precambrian but the feeder intrusions of magma, both of granite and gabbro, rising up from the subducting oceanic plate are particularly prominent as dykes seen in the northern Malvern Hills. When we walk along the Malverns we are essentially observing a deep slice of the Earth's crust uplifted by earth movements and worn down by millions of years of erosion to reveal the internal workings of a subduction zone.

Between 570 and 550Ma, the tectonic activity increased and, in particular, in the region of East Avalonia the subduction direction became oblique to the line of the subduction zone, which lay along the edge of the continent (Figs. 4.3, 4.4).

This oblique plate collision, which continued into the early Cambrian, introduced transverse stresses in the Gondwana continental margin causing it to be 'sliced up' with the slices moved laterally along tear faults. The position of these tear faults in the very different geography of today is shown in Fig. 4.4. Areas of volcanic, plutonic and sedimentary rocks were not only further metamorphosed at depth by these processes but also were moved long distances along the faults. Rocks that might have been formed many kilometres apart thus ended up next to each other. The Warren House Volcanic Formation is one of these small areas of volcanic rocks that has been moved in this way and is now adjacent to the Malverns Complex rocks. It lies just south-east of the Herefordshire Beacon.

This was the origin of the major tear faults seen today in the sub-parallel Menai Straits and Church Stretton fault systems and in the Malvern Fault, one of the principal structural elements in the Midlands, which still separates the much younger rocks of the Worcester plain from the Lower Palaeozoic rocks to the west in Herefordshire.

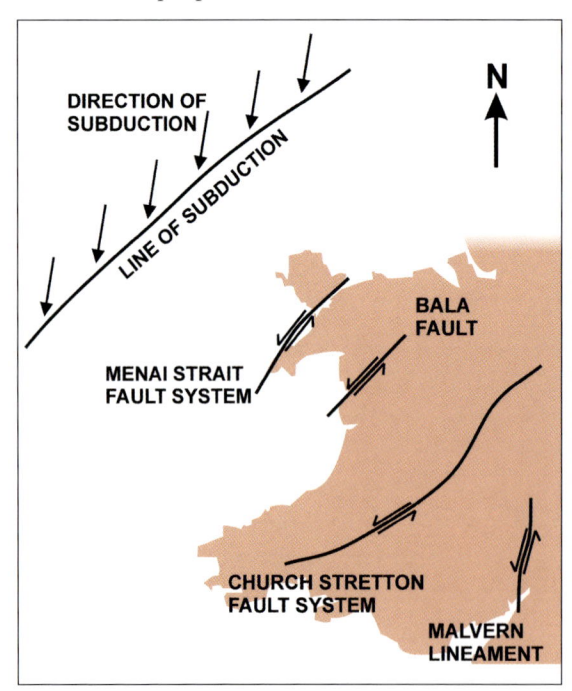

Figure 4.4 Development of the Malvern Fault and others, showing the strike-slip faulting in its location within the present land of North Wales and the Welsh Borders. The forces due to the oblique subduction fractured the adjacent crust and the fragments (terranes) moved relative to one another. (Drawn by John Payne)

The Cambrian Period – 542 to 488Ma

The Precambrian sequences of the Welsh borders were next subjected to a period of erosion leading to the eventual exposure of the rocks of the Malverns Complex at the surface. The southward motion of the Iapetus oceanic crust had virtually ceased and, consequently, so had volcanic and subduction activity. The Malvern Hills area formed part of a low-lying landscape on the borders of a widening Iapetus Ocean to the west (Fig. 3.9). Eventually, the margins of the Iapetus Ocean spread over the deeply eroded Precambrian landscape as a shallow sea probably dotted with small islands. A shallow-water sandstone, almost totally made of quartz grains, is the product of this phase of sedimentation and is locally called the Malvern Quartzite. It lies at the base of the local sequence of Cambrian rocks. The sea then became slightly deeper and fine-grained mudrocks succeeded these basal sandstones heralding the thick mud sequences laid down during the early Ordovician (Chapter 5).

The Ordovician Period – 488 to 443Ma

The earlier southward movement of the Iapetus plate was now reversed. The microplate containing the Malverns Complex and the veneer of Cambro-Ordovician sediments separated from Gondwana and was taken northward with the oceanic crust of a now-narrowing Iapetus Ocean. Eventually this small 'Avalonia' plate collided with the neighbouring microplate of Baltica in late Ordovician times. This event caused deformation which uplifted the Precambrian rocks to become islands in the succeeding shallow seas of the marine Silurian Period.[2] These were the first 'Malvern Hills' although they probably bore little resemblance to those of today. Distinctive beach conglomerates from the early Silurian can be seen in places along the western edge of the Malverns (see Chapter 6). The deformation also uplifted, tilted and eroded the Cambrian and early Ordovician sediments to produce the varying thickness of these strata along the Malvern Hills. (This episode is known as the Shelvian event after the village of Shelve in neighbouring Shropshire.) Meanwhile, the Iapetus Ocean continued to narrow and Avalonia drew ever closer to the large northern continent of Laurentia (Fig. 3.10).

The Silurian and Devonian Periods – 443 to 359Ma

The final closure of the Iapetus Ocean occurred with the collision of Avalonia/Baltica with the major continental plate of Laurentia during the late Silurian to early Devonian. There is little indication of this, the Caledonian orogeny, within the Malvern area. The area of Herefordshire and the Malverns lay to the south of the main collision zone which, to the north, gave rise to the impressive Caledonian Mountains of north Wales, the Lake District and Scotland. However, gentle uplift of the Malvern area did occur at this stage causing the sea to shallow and the marine conditions of the early Silurian Period to give way to the brackish/freshwater environments of the Old Red Sandstone (see Chapter 7).

The Carboniferous Period – 359 to 299Ma

The gradual closure of the Iapetus Ocean in Ordovician times had led to the opening of the Rheic Ocean to the south of the drifting micro-continent of Avalonia. This new ocean started to close after Avalonia collided with Laurentia (Fig. 3.11). Multiple plate collisions with Avalonia from the south caused the onset of the Variscan orogeny with its large effects on our area. This was acted out in stages from the mid-Carboniferous through to the early Permian. Significant crustal shortening (compression) and uplift were generated in south-west England and south Wales producing strong east-west structures. To the north, in the area of south-east Wales, the Forest of Dean and the Malvern Hills, a rotation of the main compression direction initiated strong north-west to south-east and north-south structures together with a series of sloping faults called thrust planes. These planes in the Malvern area follow the line of the old Malvern Fault from the Late Precambrian (Fig. 4.4), which was a continuing zone of relative weakness in the basement rocks. On this reactivated Malvern Fault, a number of thin slivers of basement rocks were thrust upwards along with the main Precambrian mass. One of these slivers forms today's Malvern Hills ridge. Figure 4.5 shows the final result of this movement following subsequent events in the Permian, discussed below.

The Variscan thrust planes, together with pre-Silurian (Shelvian) fault movements[3] which affect Cambrian, Early Ordovician and Precambrian strata, are the key elements in the current explanation of the geology on the Herefordshire side of the Malvern Hills.

As an example, Figure 4.6 shows a more detailed view of the effect of thrusting at the Herefordshire Beacon. A north-west to south-east section through the hill clearly shows the effect of westerly directed Variscan thrust planes. The Precambrian rock here rests upon the much younger Silurian rocks. The drag on the lower layers resulting from the movement of

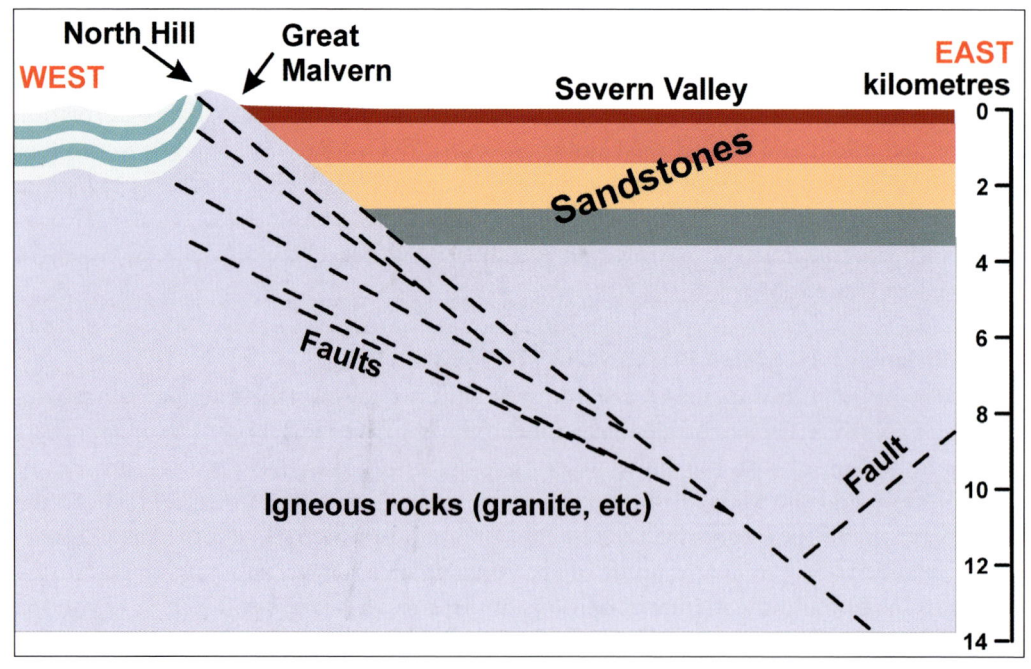

Figure 4.5 The Malvern Hills were raised by thrust on the Malvern Fault as slivers of Precambrian basement rocks, followed by 'normal' displacement of the fault resulting in the Worcester Graben. The sandstones shown in the area of the Severn Valley were deposited well after the thrusting, in the Permian and Triassic Periods and while the floor of the graben was sinking. (Drawn by John Payne)

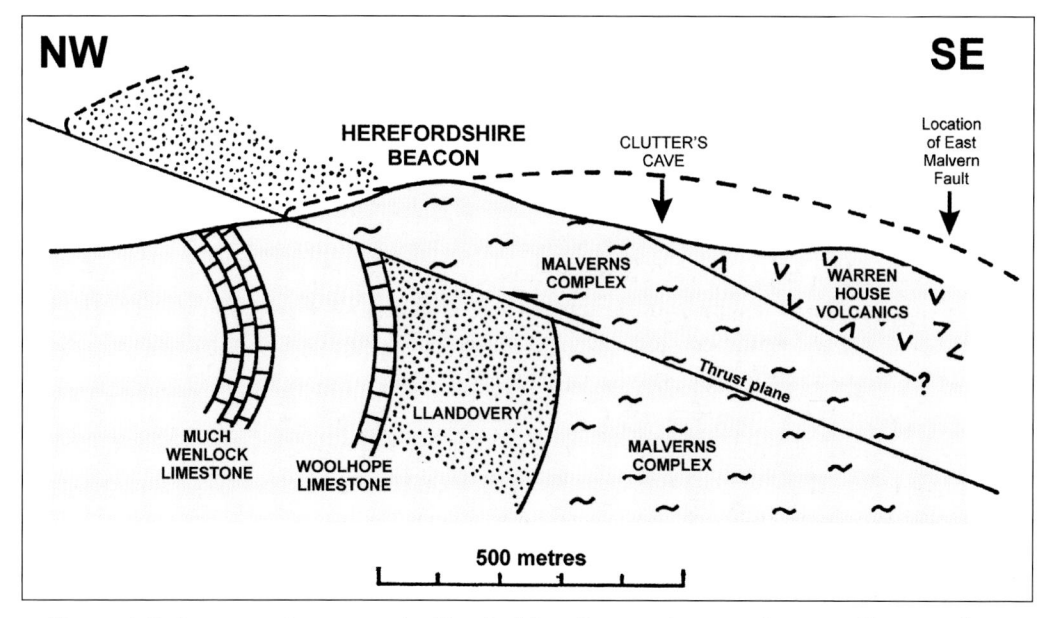

Figure 4.6 A cross-section across the Herefordshire Beacon showing the major Variscan thrust plane over steeply overturned Silurian strata. (Adapted from Fig.4 in Brooks 1970)

the Precambrian upper layer has folded and inverted the Silurian strata beneath the thrust plane. The Warren House Volcanics are seen to be transported westwards on an earlier thrust plane prior to the main Variscan movements. There is no evidence of any local Cambrian strata involved in these movements at this location.

To take this story forward, further exposures are needed, such as those to the north at Martley, detailed geophysical surveys and new borehole data.

Prior to the Variscan movement, the Silurian and early Devonian rocks to the west in Herefordshire lay as they were deposited, in horizontal layers or strata. The thrusting from the east crumpled them, giving rise to a series of folds which form major features in the present landscape of East Herefordshire, such as the Ledbury Hills and the Woolhope Dome, both of which are described in Chapter 6. The effects of the Variscan thrusting in the origin of the Malverns are discussed further in Chapter 8, particularly with reference to the structure of eastern Herefordshire.

The Permian Period – 299 to 251Ma

The mountains thrust up by Variscan movements occupied at least the area of today's Severn valley. They were largely eroded away during the succeeding Permian Period. The erosion products swept off the mountains gave rise to the Haffield Breccia of the south Malverns. An east-west extension of the crust locally, due to crustal relaxation, followed and reversed the earlier compressional regime. A further fault similar to the Malvern Fault but with westward dip was formed tens of kilometres to the east and, as these moved apart, a flat-topped prism of Precambrian rock moved downwards by about 3,000m between two essentially parallel fault lines to form a rift valley structure called the Worcester Graben.

The potential for this movement actually to result in a great rift valley seems not to have been fulfilled – as it formed, the valley was filled with erosion debris which became the strata of the late Permian and the Triassic Periods. The rock sliver of the Malvern Hills stayed in its uplifted position and now dominates the Worcester Graben and its infill of later sediments (Fig. 4.5).

Triassic Period to the Present – From 251Ma
No tectonic events have greatly disturbed the distribution of the local rocks since the Permian Period. During the desert conditions of the Triassic the Worcester Graben continued to fill and some erosion of the Malvern Hills must have occurred, although there are few clear traces of this. The Jurassic Period saw the whole area sink below sea level. The great thickness of rocks deposited in Jurassic and Cretaceous times almost certainly covered the Malverns, protecting them from erosion for 140My. This rock cover has been completely removed in the last 60My, subjecting the Hills once more to the processes of erosion. These probably became most severe only during the Ice Age of the last two million years although even then the effects seem to have been minor (Chapter 9). The Triassic rock surface has been re-exposed so that, at least nearby, on the eastern side, the essentials of the Triassic landscape have been exhumed.

The Malvern Hills: Their rocks and landscape

The Malvern Hills consist of a north-south sliver of mainly igneous rocks, up to 700My old and represent one of the largest and most important outcrops of Precambrian rocks in southern England. The relatively low southern hills of the Malverns probably contain the oldest rocks. These are now represented as a series of metamorphic schists and quartzites, originally shallow water sedimentary rocks laid down on a Precambrian continental margin, but now altered by subsequent Cadomian plate collision events.

By contrast, the impressive northern hills are dominated by diorite, an igneous rock. This rock was formed deep within the subduction zone by the solidification of a body of magma intruded into an envelope of older metamorphic crustal rocks now only seen in the southern hills. The magma resulted from the mixing of earlier basaltic, silica-poor melts with more silica-rich magmas. These diorites were themselves intruded by later granite plutons which now form the highest parts of the Malvern ridge.

Both the southern and the northern hills are intruded by a series of late coarse granite dykes which are themselves cut by a later set of microdiorite intrusions. These rocks are evidence for the further igneous activity along the subduction zone and almost certainly acted as feeders to surface volcanoes now long since eroded away.

The only truly volcanic rocks in the Malverns are the much younger Warren House Volcanics seen on Broad Down and Hangman's Hill which represent the final piece in a complex story of volcanoes, intrusion and plate collision.

The igneous rocks of the Malvern Hills
As introduced in the previous chapter, igneous rocks crystallise from a body of molten rock, or magma. Igneous rocks are divided into two main groups: ***Intrusive***, formed from magma cooling at various depths below the surface of the Earth; and ***Extrusive***, products

of volcanic eruptions, including lavas and fragmental materials, involving magma cooling at the Earth's surface. The principal ways by which magma moves to, or close to, the surface are indicated in Figs. 3.2 and 3.8, at either mid-ocean ridges or subduction zones.

The Malverns Complex is totally made up of intrusive igneous rocks. (It does not include the Warren House volcanic rocks.) Most of the rocks were formed deep within the crust. We are able to study them because the Malvern Hills have now been eroded down to their level in the crust. As the molten rock cooled under the high temperatures found at depth, crystals of a variety of minerals were able to grow from the melt. The very slow cooling allowed 'large' crystal sizes to be attained, usually greater than 1mm in length. Sometimes crystals are able, due to a high content of volatile materials and having space for crystal growth, to grow much larger and then the rock is termed a pegmatite. These deep-seated igneous rock bodies are collectively termed plutons and their environment of formation referred to as plutonic (after the Greek God of the Underworld). The various rock types are illustrated in Figure 4.8.

The rocks of the Malverns Complex are typical of those formed in island arcs and subduction zones at continental margins. This is just the situation outlined earlier for the Cadomian orogeny. They are said to be of the 'calc-alkaline' series of rocks since the minerals are rich in calcium.

Volcanoes, both single conduit and fissure types, rely on magma reaching the surface through the upper crust. These vertical structures of molten rock eventually solidify to form dykes and have a cross-cutting relationship to their surrounding rocks. Sometimes magma can penetrate along the bedding planes of sedimentary rocks beneath the volcanic area to form a sill. These structures will be concordant (i.e. parallel) with the bedding rather than discordant as in the case of dykes. Occasionally sills can transfer from one level to another in a series of small steps but essentially they remain parallel to the bedding planes (Fig. 3.18).

The solidifying rock in sills and dykes cools more slowly than lava

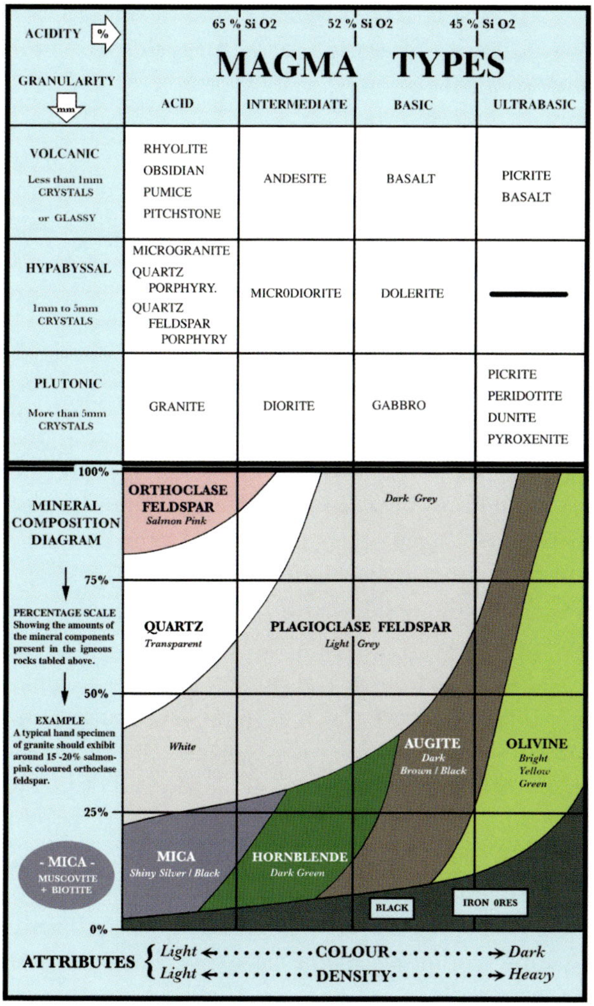

Figure 4.7 Classification of calc-alkaline igneous rocks together with their component mineral species.

(which solidifies at the Earth's surface) as it is insulated by layers of overlying rocks and thus produces crystals visible to the naked eye but generally less than 1mm in length. These crystals are said to have a medium grain size as opposed to the larger crystals seen in the associated plutonic rocks. All rocks formed in sills and dykes are collectively referred to as hypabyssal igneous rocks, a word derived from their formation at shallow levels in the crust. These minor intrusions are represented in the Malverns by a suite of dolerites and microdiorites (Figs. 4.7 and 4.8).

Good exposures of igneous intrusive rocks are seen throughout the Malvern Hills within the disused quarry workings. They are classified by their overall mineral composition which reflects the chemical composition of the original magma. Four main compositional groups are recognised.

Acid: All igneous rocks are rich in both silicon and oxygen and this is expressed by the overall percentage of silica in the resultant rock. Acid igneous rocks are richest in silica, so much so that they are characterised by free quartz (the crystalline form of silica, SiO_2). They are typically light coloured and have density of around $2.5 g/cm^3$. Feldspars, both potassium-rich orthoclase and sodium-rich plagioclase, are typically seen along with quartz and a few crystals of silver muscovite mica or black biotite mica (Figs. 4.8a and b)

The terms 'acid' and 'basic' do not have their normal chemical significance here. Their use derives from an early, erroneous geological theory that natural silica arose from a chemical reaction involving silicic acid. Thus, silica-rich rocks are termed 'acid' and silica-poor rocks are therefore termed 'basic'. The modern terms 'felsic' and 'mafic' are close but not exact equivalents. '**Fel**sic' rocks are those rich in **fel**dspar and **si**lica. '**Maf**ic' rocks are those rich in **ma**gnesium and iron (**Fe**).

The actual appearance of the individual minerals mentioned here may be found in many excellent books of mineral and fossil photographs, but fine specimens like those shown in the books are very seldom found in the field and the minerals may not be easily recognisable. The use of a small hand lens with x10 magnification is often very advantageous.

Intermediate: Formed from magma containing between 65% and 52% silica, pink orthoclase feldspars and grey translucent quartz can be seen alongside darker ferromagnesian minerals, particularly hornblende. This contrast between dark and light crystals is especially characteristic of the diorites which dominate the Malvern Hills (Fig. 4.8d).

Basic: With a much lower silica content (between 45% and 52%), free quartz is not found in this group of igneous rocks. With higher percentages of calcium, iron and magnesium in the magma a new suite of ferromagnesian minerals such as hornblende, augite and olivine is found, together with higher contents of iron oxides, such as magnetite. These new minerals, as seen in Figure 4.8f, form 'mafic' rocks and make the rock darker and of higher density than their acid equivalents.

Ultrabasic: Here ferromagnesian minerals make up the bulk of the igneous rock and, with a silica percentage of less than 45%, feldspar is often an unimportant or absent component. Small pods of pyroxenite within the Malvern Hills are a good example. Pyroxenite is an ultrabasic intrusive igneous rock mainly composed of the ferromagnesian group of minerals called pyroxenes (e.g. augite) together with minor amounts of olivine, hornblende or chromite. Like the basic rocks, these too are described as mafic.

a) Granite : Orthoclase, quartz and some mafics. Crystal size – several mm

b) Pegmatite : Orthoclase and quartz. Crystal size > 10mm

c) Diorite : Pink plagioclase and mafics. Little quartz. Crystal size – several mm

d) Diorite : White plagioclase and mafics. Little quartz. Crystal size > 1mm

e) Microdiorite : Plagioclase and mafics. Crystal size << 1mm

f) Dolerite : Plagioclase and mafics. Crystal size <<1mm

Figure 4.8 Some rock types of the Malverns Complex. Orthoclase feldspar is pink, plagioclase feldspar is white or pink, quartz is translucent grey and mafics are dark minerals. Pegmatite, microdiorite and dolerite all form minor intrusions into the main rock body. Microdiorite and dolerite appear similar but contain different mafic minerals. (The scales have divisions of 1cm.) (© John Payne)

g) Rhyolite : Orthoclase and quartz. Crystal size – microscopic. 'Flinty' fracture

h) Basalt : Plagioclase and mafics. Crystal size – microscopic

i) Schist : Shiny mica on cleavage surfaces

j) Gneiss : Minerals in evident bands

k) Epidote : Distinctive apple green mineral on surfaces

l) Altered diorite : Diorite chemically altered by metamorphism and weathering. Wide colour variation. The original crystal structure is lost, leaving an unstructured rock. Common in North Malvern walls and quarries.

Figure 4.9 Some associated rock types of the Malvern Hills.

Rhyolite and basalt are members of the Warren House Formation found on Broad Down. The other examples are metamorphosed rocks, usually from Malverns Complex originals (Box 3.3). (The scales have divisions of 1cm.) (© John Payne)

Figure 4.7 shows the classification and nomenclature of the igneous rocks. The four main magma types are shown across the top and their individual columns list the rock types which are associated with that particular magma and which crystallise in different igneous environments. Moving down the individual columns gives the main crystal species associated with each magma type and the overall colours to be seen on fresh surfaces of both medium and coarse-grained examples.

Pictures of the main rock types seen on the Malvern Hills are shown in Figures 4.8 and 4.9.

Some important sites on the Malvern Hills

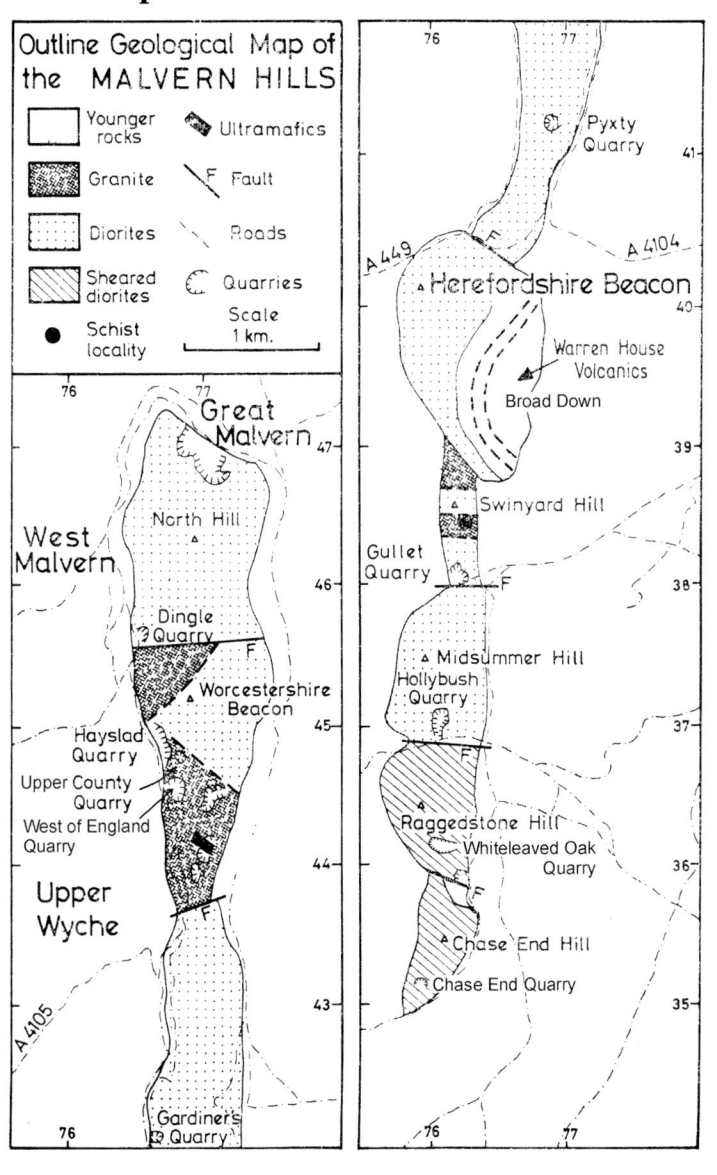

The following pages give geological details of some of the main sites on the Malverns, starting in the north and ending in the south. In this direction, as described above, the rocks suffer a general change from plutonic igneous rocks to metamorphic rocks probably derived from Precambrian sediments. The site positions are indicated in Figure 4.10.

Figure 4.10 A map showing the distribution of the main rock types in the Malvern Hills. (Adapted from Fig.1 in Blyth and Lambert 1970)

The Malverns Complex: Dingle Quarry and its geological history

The multi-coloured appearance of the walls of St James Church owes its origin to the wide range of igneous rocks (Fig. 4.7) outcropping in the Malvern ridge which rises above West Malvern. Five hundred metres south of the church, close to the Herefordshire border, lies Dingle Quarry (SO 7654 4567) where all the key events of this Precambrian igneous activity can be identified (Fig. 4.11).

On entering the quarry, it is clear that it has been worked on two levels. The predominant grey-weathered rock at both levels is diorite which, on fresh surfaces, appears distinctly granular with an equal proportion of both dark and light minerals. This 'pepper and salt' appearance is due to an intimate mixture of dark ferromagnesian (mafic) minerals, rich in iron and magnesium, such as hornblende, and lighter plagioclase and orthoclase feldspars, rich in alkali elements such as sodium and potassium (Fig. 4.7). On the north-east side (Fig. 4.11) just below the northern end of the quarry bench, a series of diffuse folds and veins of granite is seen within the diorite host (where granite pods are indicated on the photograph). Their junctions seem to merge with the diorite and suggest a process of incomplete mixing of magmas, or hybridisation, early in the history of the Malverns Complex. In the upper level towards the north, the marked pegmatite dyke is a sharp-edged coarse granite. With its large crystals of white quartz and pink orthoclase feldspar, it is designated a pegmatite by igneous petrologists (Fig. 4.8b). Here the granite shows a distinct alignment of crystals and appears almost banded, as seen in high-grade metamorphic rocks such as

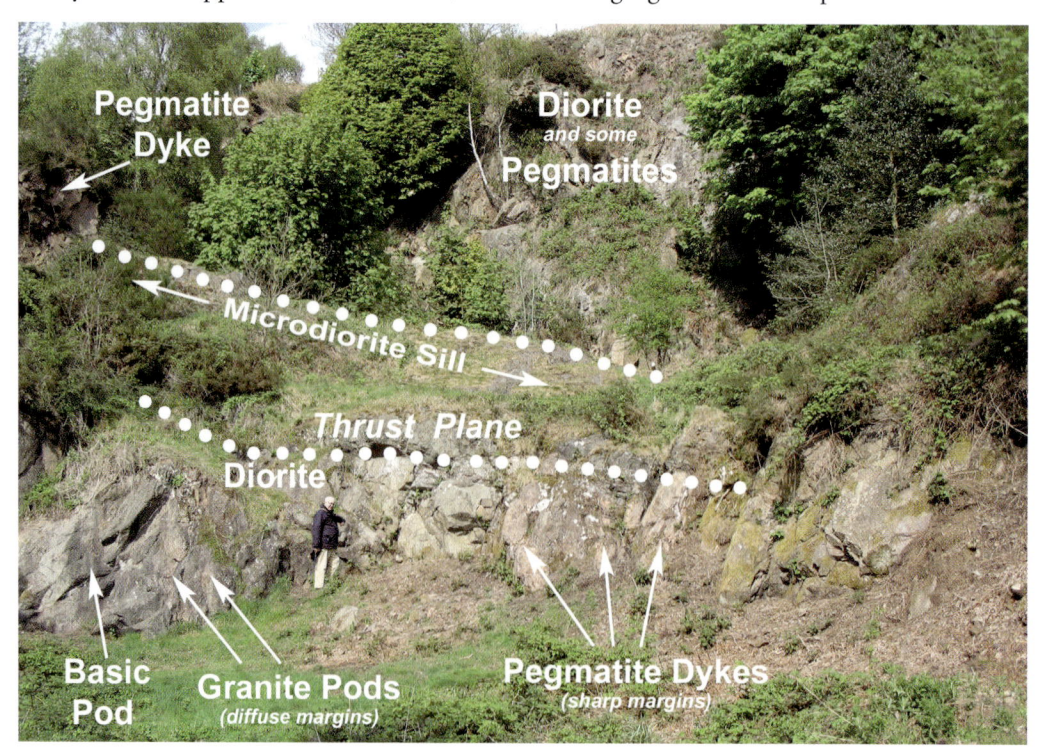

Figure 4.11 The north-east side of Dingle Quarry, West Malvern (SO 765 457) with major geological features indicated within the Precambrian Malverns Complex. (© John Stocks)

Figure 4.12 An unusual form of pegmatite with black biotite mica plus the normal constituents, clear quartz and salmon-pink orthoclase (from Lower Tolgate Quarry (SO 770 441)). (© John Stocks)

gneiss. However, this layering has been produced by local shearing effects and not, as for true gneiss, by a regional (i.e. very large-scale) metamorphic event.

The quarry of the upper level is formed of a four-metre thick body of fine-grained, dark grey microdiorite dipping 50° to the east. This sheet of rock, which is seen extending horizontally across the quarry, truncates the pegmatite dyke exposed in the lower level face and cross-cuts the foliation seen in the pegmatite margins at the upper level.[4] This intrusion is therefore one of the younger events recorded in the Malvern Hills. Often incorrectly identified as dolerite (which also possesses small-sized grains but has a somewhat different composition), these intrusions are also seen forming the conspicuous crags of Ivy Scar Rock (SO 773 464) above Great Malvern, where the flow patterns associated with their intrusion can clearly be seen low in the cliff face.

The upper contact of the microdiorite shows an irregular, stepped contact (Fig. 3.18) with the granite pegmatite and evidence of reduced crystal-size indicating a chilled margin against the older rocks. The underside of the sheet, which particularly overhangs the lower face, by contrast displays evidence of tectonic movement, in the form of slickensides, surfaces which are polished and striated due to grinding by the motion of the adjacent rock mass along the fault plane. It seems to have acted as a line of weakness in a later episode of the faulting which, although focussed on the microdiorite, has also affected the diorite host in the central zone of the lower level. Some movement of the upper level relative to the lower level has therefore occurred along the line of the microdiorite intrusion.

The Malverns Complex: South to Wyche Cutting

Moving south along the West Malvern Road from the Dingle Quarry, a group of small quarries on the lower slopes of the Malvern ridge may be visited. Hayslad Quarry, sometimes referred to as Dogleg Quarry (SO 767 449) is much overgrown but, again shows a prominent microdiorite intrusion (almost quarried away) running north to south through the host granite. As in Dingle, this intrusion is associated with a major phase of north-south faulting and brecciation.

Upper County Quarry (SO 768 448) was once quarried for its dolerite dyke, a late intrusion showing no evidence of shearing but containing several enclaves of the earlier granitic host. Just south of this quarry and lower on the slopes, the larger lower part of the West of England Quarry (SO 7677 4468) shows the whole range of the Malverns Complex with both granites and diorites of varying grain sizes dominating its crags. Close to these quarries is Hayslad Spring (SO 766 448) which originally lay further uphill and is now a roadside spout fed from covered reservoirs capturing the natural spring flow. It is often crowded with visitors and locals filling water containers.

Further south, we enter the Wyche cutting (SO 769 437) which sits directly on the Herefordshire border. This narrow road cutting, where all geologists need to be particularly alert for traffic, lies on a fault zone between metamorphic schists on its the south side and more massive granite on the north side. The quartz-mica-schists show a very variable foliation direction. This seems to be the result of intense deformation involving movement from the south-south-east, which was deflected around the massive, largely granite, plutonic mass of the north Malverns. Brammall[5] designated this thrust episode rather colourfully as the 'Cheltenham Drive' alluding to the direction of origin of the initial movement.

One of the main issues yet to be fully resolved in the Malverns is the origin of rocks resembling schists and gneisses (Box 3.3). These would normally be the result of sustained heat and pressure at depth within mountain belts; in other words, a product of regional metamorphism. However, in the Malverns, localised intense deformation, or shearing, has occurred during the various orogenies. This has resulted in a mineral fabric which closely resembles that of schists and gneisses and it led to much confusion amongst 19th-century geologists. The shearing causes the alteration of the original minerals in the granites and diorites, such as hornblende and biotite mica, to minerals that are stable under this shearing regime. Such minerals are pistachio-green epidote, dark green chlorite mica and silver muscovite mica all of which have already featured in the quarries of the north Malverns. These episodes produce localised areas of schistose and gneissose rocks, much more restricted in outcrop than normal regional metamorphic products and containing a very limited number of metamorphic minerals. Callaway[6] designated this process as 'dynamic metamorphism'.

The Malverns Complex: The Herefordshire Beacon
Highly foliated quartz-mica-schists form the ridge just south of the Wyche (Fig. 4.13). The Colwall Railway tunnel was driven through the Malvern Ridge at this point by an army of Welsh and Irish navvies armed only with picks, shovels and black powder. Altered diorites were the main rocks encountered at depth. The workers endured tremendous hardship in tackling some of the hardest rocks in England. It was also a problem financially, with two firms going bankrupt before the first tunnel was completed in 1861 under the directions of Stephen Ballard, a local engineer. Subsequently, minor collapses in the tunnel in 1907 led to the building of a second tunnel, opened in 1926, whose deep geology was reported by Robertson.[7]

Altered and sheared diorites, with considerable injection of diffuse veins of granite material, are seen in the south face at Gardiners Quarry (SO 7660 4210). In the southern part of the quarry, a prominent shear zone is thought to represent a continuation of the Colwall Fault which affects both Palaeozoic and Precambrian rocks in this area.[8]

Our route south now takes us to the Malvern Hills Hotel and the large car park on the A449. Here, as at the Wyche, there is a major north-west to south-east transcurrent fault running across the Malvern ridge. The effects of this are clearly seen if one stands on the

Figure 4.13 (above) Looking south-west across the middle part of the Malvern ridge. Malvern Wells lies in the foreground and the Herefordshire Beacon to the left with the wooded curved limestone ridge of the Ridgeway behind it. The photo shows the steep eastern side of the Malvern Hills. Essentially, this steep slope is the fault plane of the East Malvern Fault, on which there was about 2500m of movement in Permian and Triassic times. Close to the far side of the Malvern ridge, the wooded edges of Silurian limestone strata can be seen. These dip beneath the flat valley floor before rising again in the distant Ledbury hills. (© Derek Foxton Archive)

Figure 4.14 (below) The Iron Age hillfort (British Camp) on the Herefordshire Beacon rises above the reservoir, completed in 1894 for Malvern's water supply. The curved form of the ridges and vales beyond the hill reveals an anticline plunging to the right (north).(© Derek Foxton Archive)

Herefordshire Beacon (338m) and looks directly north (Fig. 2.2). The westward displacement of the Herefordshire Beacon from the line of the ridge to the north is evident. The fault forms the northern end of the Beacon, which is displaced to the west by about 1km on an underlying low-angle fracture or thrust (Fig. 4.6).

The dip of this important thrust plane is between 16° and 18° to the east or southeast.[9] The Precambrian thrust block is known to over-ride inverted Silurian Woolhope Limestone near Walm's Well (SO 7605 3926), north-north-west of Swinyard Hill. At its western margin the thrust block rests on inverted or vertical shales of the Coalbrookdale Formation, close to their outcrop with the Much Wenlock Limestone, here seen dipping at 75° to the east. It is thought that the Warren House Volcanic rocks, which outcrop on Broad Down (SO 764 395), constitute another similar thrust block pushed westwards onto the Malverns Complex (see later section).

The Malverns Complex: Metamorphism on Swinyard Hill
Beyond Broad Down and the 'Silurian Pass' (SO 762 390), first named by Phillips[10] after its early Silurian (Llandovery) outcrops and the location of another fault crossing the ridge, the ridge rises again towards Swinyard Hill (SO 762 386). The Malverns represent not only a complex suite of igneous rocks intruded in various phases, but also a complicated history of subsequent metamorphism. Here on Swinyard Hill, evidence of the metamorphic events is perhaps best displayed. The northern peak of pink biotite-rich granite gives way southwards to coarse dark-green hornblende-rich rocks (amphibolites). These are products of a high pressure and high temperature alteration of basic igneous rocks, shot through with a network of thin granite veins and clumps of lustrous 'books' of muscovite mica crystals. Further metamorphic rocks, irregularly banded and containing alternating layers of pale feldspars and quartz and darker green chlorite mica on a millimetre scale are also seen and have been identified as sheared gneisses. Smaller outcrops of garnet-bearing mica schists may represent slivers of the former sedimentary cover into which the Precambrian plutons were intruded. If so, they are the oldest rocks on the hills and must go back to at least 700Ma. It is likely that these are the result of very old regional metamorphism, which may generate garnets, rather than the more recent dynamic metamorphism, due to shearing events, which affects most of the Malverns (Box 3.3; Fig. 4.15).[11]

South of Swinyard Hill, Gullet Quarry (SO 762 381) cuts right across the Precambrian ridge. This was one of the most famous quarries in the Malverns, along with North Quarry, and was the last to be worked. With its spring-fed lake and landscaping by the Malvern Hills Conservators, an important natural-looking feature has been produced out of the former desolation.

The high vertical faces on the north side of the quarry consist mainly of massive diorites cut by a wide variety of dykes and late muscovite-rich pegmatites. Intense shearing and fracturing of all rock types, except the late pegmatites, on both a small scale and along major shear planes is the most striking feature of this complete section through the Malverns ridge (Fig. 4.16). This quarry is dangerous and on no account should the rock faces be approached. A distant and general view, though now largely obscured by vegetation, may be had from the designated footpath which skirts around the south of the quarry.

Figure 4.15 Gullet Quarry, from the south-east, shows the position of the Gullet Pass, a valley which follows the line of a geological fault running across the Malvern ridge. This is one of only two valleys cut across the line of the hills and its stream rises to the west on the Herefordshire side, crosses the Malvern Axis and flows down to the River Severn. The quarry is dug into the end of Swinyard Hill which leads northward to the Herefordshire Beacon behind with its Iron Age hillfort. The intermediate ground, the lower hilly area to the right, is Broad Down, an area of volcanic rocks (the Warren House Formation). The nearest hill, in the foreground, is Midsummer Hill, the location of a second hillfort, hidden in the trees. (© Derek Foxton Archive)

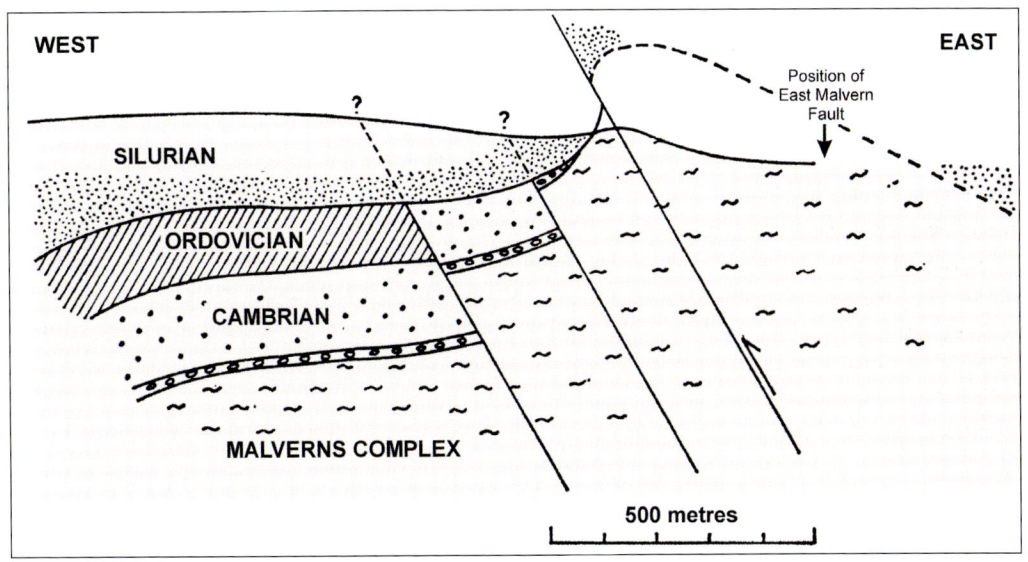

Figure 4.16 A cross-section across the Malvern Hills in the area around Gullet Quarry showing the relationship between the Malverns Complex and the unconformable Lower Palaeozoic strata. (The figure shows the situation prior to the Triassic downfaulting to the east of the hill.) (Adapted from Fig.3 in Brooks 1970)

The Malverns Complex: The Southern Hills

Moving south from the Gullet along the Worcestershire Way takes us onto Midsummer Hill (284m; SO 760 375) and its Iron Age hillfort. From here, there are good views of Eastnor Castle to the west, to the east across the Worcestershire Basin to Bredon Hill and the Cotswolds as well as a view back to the north along the Precambrian ridge.

As we descend, gently at first but soon very steeply, the massive excavation of Hollybush Quarry comes into view. It was closed in 1977 and is now flooded. To its west, on the Herefordshire side, is Slasher's Quarry (SO 759 371), now used as a rifle range with restricted access. The wide range of Malverns Complex rocks in this quarry includes hornblende-rich examples. These represent early melts from the original dioritic magma. They are similar to the rocks of the north Malverns and contrast strongly with the exposures on Swinyard Hill. The concept of strongly individual blocks of Malverns Complex separated by transcurrent faults and moved by varying amounts to the west along Variscan thrust planes (Fig. 4.10) is therefore well illustrated in this area.

The roads crossing the Malvern ridge all take advantage of the enhanced erosion which has generated low points where the tectonic dislocations cross the line of the hills. The A438 runs through the Hollybush pass and, on its south side, the summit of Raggedstone Hill (254 m; SO 760 364) offers magnificent views of Herefordshire to the west.

The double peak which gives the hill its name is due to a Variscan fracture which cuts the hill from north to south and has been preferentially eroded out between the two summits. This has allowed a thin outcrop of basal Cambrian quartzite to be preserved, wedged between the Precambrian masses but seen only in the valley on the north side of the hill. The quartzite appears in similar situations on Midsummer Hill (see Chapter 5).

Following the ridge southwards, to the east there is a remnant of an early boundary or defensive embankment cut out of the solid Malverns Complex and called the Red Earl's Dyke. This is seen along much of the Malvern ridge today and is significant in forming the Herefordshire border. A steep descent on a path through gorse and woodland leads to the secluded hamlet of Whiteleaved Oak. Here, the three counties of Herefordshire, Worcestershire and Gloucestershire all meet. On the eastern side of the hill, hidden in woodland, is Whiteleaved Oak Quarry (SO 761 360).

This quarry represents the best exposure of metamorphic rocks in the Malverns (Box 3.3). Localised intense deformation, or shearing, has resulted in a series of rocks that closely resemble schists and, more rarely, gneisses. The shearing has altered the original igneous minerals of the diorites, such as hornblende and biotite, to minerals that are stable under the extreme heat and pressure which were generated in the semi-plastic Malverns Complex at depth beneath the rising Cadomian mountains.

The lamination produced by the new minerals such as epidote (pale green), chlorite (dark green) and muscovite (silver) formed by these processes was initially erroneously interpreted as the typical gneissose and schistose banding of a series of regionally metamorphosed rocks. The clue to their real origin in a Precambrian shearing event lies firstly in their very localised occurrence and secondly, in the breakdown of the high temperature igneous mineral species to a low temperature suite of epidote, calcite and muscovite.

Much of the diorite base has been reduced to a very fine-grained, silvery rock by this Precambrian shearing event. This is seen in the scree on the south side of the quarry. Occasional pegmatites show less deformation, but are usually stretched and thinned until rupture occurs, a process called boudinage. This causes small lenses of coarse granite to seem to float in the silvery schist matrix of the quarry faces. The pegmatites indicate that their intrusion was probably due to a second magmatic episode at depth and later than the earliest shearing episode. However, the boudinage seen suggests that further shearing episodes continued beyond the date of their intrusion. Despite this overall emphasis on shearing in the Malverns, there is some evidence of an earlier regional metamorphic event best seen at Whiteleaved Oak by the growth of small garnets within the more massive quartz-rich areas of the quarry faces. These may represent the metamorphism of highly silicic pegmatites or sandstones within the pre-Malverns Complex sedimentary envelope. (Note that the north side of the lower of the two levels of this quarry is overhung by a large tree rooted in faulted rock and so must be regarded as particularly unsafe.)

South of Whiteleaved Oak, sitting proudly in Gloucestershire, is the final outcrop of the Malverns Complex, at Chase End Hill (191m; SO 761 355). Sometimes referred to as the Gloucestershire Beacon, its name derives from its position at the southern end of the Royal Forest of Malvern Chase set up by William the Conqueror. Chase End Quarry (SO 758 350), south-west of the summit, shows excellent further evidence of shearing and boudinaged granite pegmatites. The late major shear planes and thrusts are perhaps better seen here than at Whiteleaved Oak, particularly in the low eastern face of the quarry.

Warren House Formation

Although the earliest Malvern volcanoes and their products are lost to erosion, a younger Precambrian group of volcanic rocks is seen to the east and south-east of the Herefordshire Beacon. This is the Warren House Formation and is quite different to the Malverns Complex. Similar small outcrops of these Precambrian rocks occur throughout the Midlands, such as at Charnwood north of Leicester and in south Shropshire, each with its own distinct suite of igneous rocks. They share a common history of formation within an extensive late Precambrian island arc but were originally widely separated along its length. As described in the section on the evolution of the Malvern Hills, earth movements around 560Ma involving fracturing and sideways movement close to the subduction zone brought together these 'slivers' of very different rocks. The Warren House Volcanics represent one of these Precambrian units now in juxtaposition with the much earlier Malverns Complex.

The pillow lavas outcropping at Clutter's Cave (SO 7628 3935) are part of a broader outcrop of volcanic rocks occupying Tinkers Hill, Broad Down and Hangman's Hill to the east of the Herefordshire Beacon (Figs. 4.6 and 4.10). The basalts at Clutter's Cave are joined by younger, easterly-dipping volcanic rocks including rhyolite lava flows, some erupted underwater, and ignimbrites, which are the solidified deposits from pyroclastic flows generated by the most explosive eruptions. All these types can be seen outcropping south of the British Camp Reservoir. They are the only extrusive igneous rocks in the Malvern Hills. Rhyolite and basalt are pictured in Figure 4.9. All the volcanic rocks are intruded by dolerite dykes and show fewer signs of deformation than the adjacent Malverns Complex.

BOX 4.1   **Clutter's Cave**

Situated on the ridge path where the easterly sloping Broad Down meets the axis of the Malvern Hills, this hand-hewn cave (SO 7628 3935) was thought to be a medieval hermit's dwelling. Some of the ideas based on that conception are given below. However, there are no records of the cave earlier than the mid-19th century and it is probably a grotto within Eastnor Park.

The cave is also known as the Giant's Cave or Walm's Cave; the latter name derives from a spring called Walm's Well (now enclosed) which lies below the cave on the boundary of News Wood. This could have been one of the springs supplying water to the nearby British Camp. Smith in his *History of Malvern* notes that from the cave the sun can be seen to set on Midsummer's day over the Arthur's Stone Neolithic chambered tomb near Dorstone (SO 319 431). Certainly a very clear day and superb eyesight would be needed to observe the phenomenon! Another issue, given the Earth's precession over the centuries that have elapsed, would be whether the observation was valid during Neolithic and Bronze Age times.

On the slope below the cave lies a large block of igneous rock from the Malverns Complex (SO 7615 3935). Referred to as the Sacrificial Stone in the Woolhope Club *Transactions* of 1889, it is thought to be the same as the Slew Stone named as a boundary marker in older documents. A local tale is that it is 'the door of the Giant's Cave thrown down' and its connection with sacrifice was enhanced by the observation that the Midsummer Day's Sun rising over the stone's surface highlights the stone's shape which fits the reclining human back and could have facilitated such rituals. Alfred Watkins, President of the Woolhope Club in 1919, developed these ideas into a ley line which included Woolhope Church, Aconbury Camp and the Gospel Oak on the Ridgeway.

These orientations, expanded in his book *The Old Straight Track*,[12] are now thought to be largely coincidental with only Aconbury Camp being of the same age (i.e. Iron Age) as the time of possible usage.

Figure 4.17 Clutter's Cave (SO 7628 3935) – a small cave (right) displaying spilitic pillow lavas (left) from the Warren House Formation. (© John Stocks)

Structurally, the thrust fault contact between the volcanic rocks and the underlying Malverns Complex dips east at about 35°. This is believed to be truncated at depth by the Herefordshire Beacon Thrust (Fig. 4.6).

At Clutter's Cave (Fig. 4.17; Box 4.1), the purple to dark grey basalts are strongly vesicular. Vesicles are spherical cavities found in lavas (and some shallow intrusions) as the result of gases dissolved in the melt being released to form bubbles when the pressure is reduced on reaching the surface. Some of the vesicles in the individual pillow margins contain needle-like crystals of epidote. The pillow structure shows that they were erupted under the sea. (The very rapid cooling of the extruded lava in the seawater leads to the formation of a hard crust around a still molten interior and the generation of successive pillow-shaped 'blobs' resting on one another.) The underwater environment led to the basalts being chemically altered and enriched in sodium derived from the seawater. Plagioclase feldspars such as the sodium-rich albite have replaced the calcium-rich anorthite typical of normal basalt lavas and led to a release of both calcium and silica. The calcite veining and epidote-rich amygdales seen at Clutter's Cave are a direct result of this process. Amygdales are vesicles which have later become filled with other minerals, commonly calcite or quartz.

In terms of age, the zircons in a crystal lithic tuff from Broad Down have given an eruption age of 566Ma (uranium-lead dating).[13] Pebbles of similar volcanic rocks have been found in the Lower Cambrian Malvern Quartzite confirming their age as Precambrian.

Thorpe[14] suggested that the overall geochemistry of the volcanic rocks indicates an island arc origin and concluded that the basalts could be an 'ocean floor' type. Certainly, the occurrence of acid lavas and pyroclastic flow deposits supports a former island arc environment, perhaps within the back-arc basin (Box 3.2).

Precambrian rocks in North-West Herefordshire

The southerly extension of the Church Stretton Fault (Figs. 4.4 and 6.5), part of the more extensive Welsh Borderland Fault system, runs along the English-Welsh border beyond Kington. Three fault-bounded hogback hills, Hanter Hill (SO 252 570), Worsell Wood (SO 258 577) and Stanner Rocks (Fig. 2.10; SO 263 585) lie within this zone. (They are just outside Herefordshire.) These hills expose a series of Precambrian igneous rocks including dolerite, gabbro and minor granitic types. A date of 702Ma (rubidium-strontium dating) confirms this suite of rocks as the oldest in southern Britain.[15]

The Stanner-Hanter Complex is grouped with the Malverns Complex as part of the earliest phases of the Cadomian island arc constructed around 700Ma, described earlier in this chapter.

Dark coloured grits and conglomerates also occur as faulted inliers close to Old Radnor (SO 250 591) on the Herefordshire border. These were exploited for road aggregate (Figs. 4.18 and 4.19) and have been correlated on lithological grounds with the late Precambrian Longmyndian rocks of the Church Stretton area.[16] The conglomerates contain pebbles identical to Stanner-Hanter intrusive rocks, supporting this Precambrian date.

Further Longmyndian rocks outcrop in the Pedwardine area, south-west of Brampton Bryan village.[17] This Brampton Formation, made up of purple-green and reddish grits, dips 60° to 70° north-north-west within Brampton Bryan Park (SO 366 718). The rocks

Figure 4.18 Gore Quarry seen from the north, a source of Precambrian sedimentary rock. This area shows spectacular scenery which has developed along the line of the Church Stretton Fault System, including most of the area of the Precambrian inliers in eastern Radnorshire. Just in the frame on the left side is Stanner Rocks. To its right are Worsell Wood and then Hanter Hill. Behind these is Hergest Ridge. The valleys running to the south-east between these hills are occupied by branches of the Church Stretton Fault System. (© Derek Foxton Archive)

Figure 4.19 Looking towards the east, Dolyhir Quarry is in the centre of the foreground with Strinds Quarry to its right. Both are quarried for Precambrian sedimentary rock. Crossing the frame in the near distance is the line of Precambrian igneous rocks forming Stanner Rocks, Worsell Wood and Hanter Hill, which were intruded along the line of the Church Stretton Fault. Beyond that, in the centre, is the valley of the Back Brook between Bradnor Hill and Hergest Ridge, leading to Kington. This valley was occupied by a glacier during the Ice Age and was a meltwater channel afterwards. (© Derek Foxton Archive)

are partially conglomeratic with vein quartz, rhyolites and schists occurring as clasts. These exposures are not accessible as they are on private land. However, as they were quarried for the local building stone, they can be examined in the walls of the nave of the nearby church of St Barnabas.

The Longmyndian rocks at Pedwardine (SO 365 708) have been thrust eastwards over Lower Ordovician (Tremadoc) rocks. A similar situation exists in the Hopesay area of Shropshire to the north. This Pedwardine inlier, bounded by two north-north-east-trending faults, is part of the same Church Stretton Fault Zone as the better exposed Old Radnor inlier discussed above.

Windows onto the Precambrian

North along the Malvern axis at Whippets Brook (SO 761 480) is a lens of faulted Malverns Complex within the core of the Storridge Anticline, a part of the anticlinal structure formed by the raising of the Malvern ridge in the Carboniferous Period (Fig. 8.13). Geophysical studies in this area find the Precambrian basement to be at a depth of 300m. As with the Gullet Pass exposure (see Chapter 5), limited Cambrian sedimentary thicknesses are indicated by a small fault-bounded lens of Malvern Quartzite in the Cowleigh Park area (SO 763 476).

The most northerly outcrop of the Malverns Complex, 15km north of the main Malvern Hills ridge, is at Martley, close to the county border in Worcestershire. A small aggregates pit (SO 745 596) since early Victorian times and first described by the eminent geologist Roderick Murchison,[18] this key location continued to be described throughout the 19th and 20th centuries. Each geologist added further observations and hypotheses to explain this extremely complex zone straddling the structurally important Malvern Fault.

By the 1990s, the pit had become completely filled in and a stand of young fir trees marked its former location. The full structural complexity has only been discovered since early 2010, when the Woolhope Naturalists' Field Club and the local Teme Valley Geological Society spearheaded the digging of a series of temporary trenches. The inlier comprises Precambrian Malverns Complex and Cambrian Malvern Quartzite surrounded by Halesowen Formation (Carboniferous). These strata unconformably overlie the Upper Silurian Moor Cliffs Formation (previously known as the Raglan Mudstone Formation).

In terms of the Precambrian/Cambrian relationships, the Malvern Quartzite at Martley is faulted against the sheared diorites (Fig. 4.20) and pegmatites of the Malverns Complex along a 5m-wide fault zone. This zone is mainly steeply dipping but, at one location, the Precambrian is seen to be thrust over the Cambrian from the west. Seismic interpretation

Figure 4.20 Malverns Complex diorite from Martley Rock (SO 7450 5956). (© John Stocks

suggests that the Precambrian at Martley has been uplifted by about 1,000m from its deep basement position.

This Woolhope Club research finished in 2015 with more temporary trenches being dug within the inlier. A paper has been published in the Club's *Transactions*.[19] Two of the original trenches have been preserved and observation platforms and information boards installed as part of a local geological trail (Teme Valley Geological Society *Martley Rock Trail Guide* (2012).)

Further reading
A good introduction to the scenery of the Malvern Hills is found in *Walks around the Malverns* by Roy Woodcock and published by Meridian Books (2002). Within this pocket-sized book are a series of walks starting from different points on the hills including not only some of the varied geology but also the historical background.

For more detailed itineraries, the two trail guides on the Malvern Hills published by the Herefordshire and Worcestershire Earth Heritage Trust are recommended. (See 'EHT and BGS Publications', pages *xix-xx*.)

Also see the Geological Conservation Review, vol. 20 (2000) *Precambrian Rocks of England and Wales*, Carney, J.N. *et al*.

5 THE CAMBRIAN AND ORDOVICIAN PERIODS

Early life in the shallow seas

Events during the Cambrian Period were critical to the evolution of life on Earth. The base of the Cambrian system of rocks is defined by the appearance of the first abundant fossils of multi-celled animals (metazoans). Although calcium phosphate is still present in many of the invertebrate shells and hard parts at this time, there is an increasing use of calcium carbonate, suggesting a rise in dissolved carbonates in the early Cambrian seas. This resulted in an explosion of life with the appearance of most of the known marine invertebrate groups (or phyla) such as arthropods, molluscs and brachiopods.

Several other key factors can also be identified at this important time in Earth's history. Firstly, the construction of the supercontinent of Gondwana in the late Precambrian through a series of major subduction events and their associated continental volcanic arcs (see Box 3.2) led to higher levels of carbon dioxide in the atmosphere. The planet warmed substantially at this stage, the remnant ice sheets of the major late-Precambrian glaciation (known as 'Snowball Earth') retreated and the sea levels began to rise.[1] Recent evidence suggests that there were no continents at the poles and that a uniformly warm climate existed on the predominantly equatorial landmasses. The rising seas inundated the margins of these Precambrian continental fragments giving rise to a new set of ecological niches in which life could diversify. Secondly, a reduction in the numbers of oxygen-depleting bacteria at this time led to higher levels of dissolved oxygen in these newly expanded seas alongside the steady rise in dissolved lime content discussed above. All these factors contributed significantly to an increase in biodiversity.

In terms of the local area, 'Herefordshire' formed part of a micro-continent (Avalonia) at a latitude of about 60°S and close to the north-west edge of the Gondwana supercontinent. The county lay immediately to the south-east of the Welsh Basin. The Basin was bounded by the Menai Straits Fault Zone and the Welsh Borderland Fracture Zone which includes the major Church Stretton fault line which runs just outside north-west Herefordshire (Figs. 4.4 and 5.1). The defined Welsh basin was probably about 40% wider than indicated by the present geology and had shallow marine platforms at its margins with deeper water sediments being laid down over much of what is now the Welsh landmass. In terms of latitude, Avalonia was not one of the tropical or equatorial landmasses and therefore is expected to show a lower level of biodiversity. This may explain the relative scarcity of trilobites, brachiopods and other invertebrates within the local Herefordshire successions.

At the end of the Precambrian and before the deposition of the earliest Cambrian sediments, there was an interval during which the older rocks were subjected to erosion and were worn down to an irregular low-lying surface. It was across this surface that the Cambrian sea advanced and, in Herefordshire and Shropshire, laid down a thin sequence of shallow marine sediments (Fig. 5.1). As the sea advanced across the area, shallow water sediments gave way to deeper water types. At any particular location, therefore, deeper water shales were eventually laid down upon earlier, shallow-water conglomerates and grits, as seen in the local Cambrian rock sequence.

Rocks of Cambrian and Ordovician age are exposed in only three small areas in Herefordshire. The largest of these is on and immediately to the west of the south end of the Malvern Hills. Here, the younger rocks capping a northward-plunging anticline (suggested

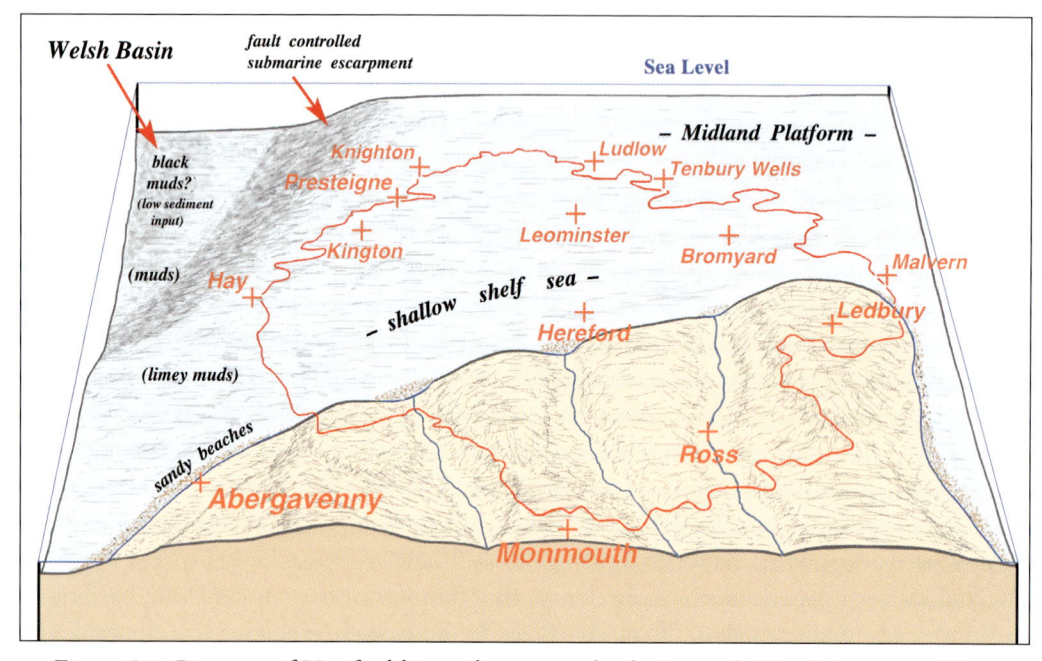

Figure 5.1 Diorama of Herefordshire palaeogeography during early Cambrian times (about 530Ma). This diagram shows the northern part of a low lying land area, composed of late Precambrian volcanic rocks and sediments, worn down by erosion over 20 million years since the last volcanic outburst (566Ma; Warren House Formation). Known by geologists as the Midland Platform, a large part of this area had already been drowned by the early Cambrian world-wide rise of sea level, to produce a wide shallow water shelf to the north, west and east of the land area. A fringing quartz sand beach (Malvern Quartzite) indicates extensive chemical weathering of the volcanic rocks. Finer sediments and clear water limestones were deposited offshore. A lack of Cambrian rocks, especially in boreholes, indicates the position of the land area. The influence of the Malvern fault system is already seen to the east, controlling the North–South coastline, and would develop further into a deep trough on the site of the present Severn Vale, later in the Cambrian and into the lowest Ordovician. To the north-west, the Welsh Basin was already subsiding down the line of the Church Stretton Fault System.
(Drawn by Gerry Calderbank from a conceptual sketch by Dave Green)

by the 'Ridgeway' exposure of Silurian limestone shown in Fig. 2.3, which shows an area slightly to the north) have been removed by erosion, creating a window about 3km long and 1km wide onto these old rocks. Elsewhere, a tiny exposure generated by faulting is found near North Malvern and movement on the Church Stretton fault system has yielded a small area in north-west Herefordshire. Exposures of Cambrian rocks are uncommon in England.

The Cambrian rocks of Herefordshire

The oldest Cambrian rock, the Malvern Quartzite, is best exposed in the south of the Malvern Hills. The rock consists of well-cemented conglomerates and sands lithified into pale grey quartzites which represent deposits left as shingle and sand on an ancient beach. Many of the included pebbles can be matched with similar igneous rocks within both the Malverns Complex and the Warren House Volcanics.[2] This constituted the earliest proof that the Malvern rocks indeed originated before the Cambrian Period.[3]

Both at Gullet Pass (SO 7597 3799) (Fig. 5.2) and at Hollybush Middle Quarry (SO 7597 3711) (Fig. 5.3) the Malvern Quartzite is seen to lie unconformably on the Precambrian. At least five beds of alternating conglomerate and quartzite appear to be

Figure 5.2 The Gullet Pass pit following clearance by the Earth Heritage Trust in 2014, showing the bedding planes of the Malvern Quartzite. The inset shows trace fossils of trails on the Cambrian beach. (© John Payne)

Figure 5.3 The Cambrian-Precambrian unconformity at Middle Hollybush Quarry with Malvern Quartzite above and Malverns Complex rocks below. (© John Payne)

draped over the Precambrian at the latter locality.[4] Fragments of horny inarticulate brachiopods such as *Micromitra sp.* characterise a sparse fauna.

Further small outcrops of these quartzites are seen in a fault-bounded lens in the Cowleigh Park area to the north-west of the hills and at Martley in nearby west Worcestershire where highly sheared, yellowish quartzites of the same age are characteristic of this structurally complex inlier (see Chapter 4).

The Lower Cambrian Hollybush Sandstone overlies the Malvern Quartzite. It is exposed in small old quarries at Whiteleaved Oak (SO 7617 3584) and on the south side of the main A438 as it approaches the summit of the Hollybush pass from the west (SO 757 368) (Fig. 5.4). Its lower layers are flaggy and micaceous, whilst above are greenish sandstones. The distinctive green colour is due to an abundance of both glauconite and chlorite. (Glauconite is a green potassium iron silicate, closely related to micas and common in marine sediments such as greensands.) Ripple marks, trails, burrows and the presence of glauconite all point to a shallow-water, marine environment which developed as the sea extended over the land surface.

The overall outcrop of the Hollybush Sandstone stretches from just south of the Gullet along the western side of the Malvern Hills to the hamlet of Whiteleaved Oak. At the top

Figure 5.4 Lower Cambrian Hollybush Sandstone outcropping on the south side of the A438 as it crosses Hollybush Pass (SO 757 368). (© Adrian Wyatt)

Figure 5.5 Hollybush Sandstone (to the left of the hammer) lying unconformably on Malverns Complex on the north-western edge of Whiteleaved Oak Quarry (SO 761 360). (© John Stocks)

Figure 5.6 Dark grey Whiteleaved Oak Shales (Upper Cambrian) beneath hedgerow at Whiteleaved Oak (SO 7582 3615). (© John Stocks)

of Whiteleaved Oak Quarry (SO 7606 3604), just below the path leading to the summit of Raggedstone Hill, the sandstone is seen to be filling shallow channels in the Precambrian and is characterised by the development of conglomerate lenses (Fig. 5.5). Inarticulate brachiopods together with hyolithids are the main fauna. (Hyolithids are an extinct group of shellfish probably related to molluscs. Their phosphatised cone-like shells are commonly found in early Cambrian strata worldwide.)

After minor Mid-Cambrian earth movements and erosion, a quieter, deeper water environment saw the deposition of black fissile mudstones of Upper Cambrian age as indicated by their trilobite fauna and the occurrence of crustacean *Cyclotron lapworthi*. The Whiteleaved Oak Shales give rise to a series of poor outcrops, notably beneath overhanging hedgerows, trending north-west to south-east through the hamlet (e.g. SO 7582 3615 and Fig. 5.6). These Cambrian shales and their similar Ordovician successors floor the vale east of the Malverns near the hamlet of Whiteleaved Oak, partly shown in the lower left of Figure 4.1.

These Upper Cambrian sediments show, worldwide, a significant dip in biodiversity. This has been associated with the continued rise in carbon dioxide in the atmosphere due to subduction-related volcanism and the increasingly high temperatures. These extreme 'hothouse' conditions initiated mass extinctions within some major invertebrate groups and an increased acidification of the oceans.

While Herefordshire and the Midlands were relatively stable in the Cambrian and accumulated a thin veneer of sediments, the area to the north-west in what is now the Rhinog mountains of the Harlech Dome was a deep basin (known as the 'Welsh Basin') in which great thicknesses of sands and muds were deposited. This contrast between deep-water sediments to the west and shallower-water continental-shelf deposits in Herefordshire and south Shropshire remains a constant feature of our geological history throughout the subsequent Ordovician and Silurian Periods, testifying to the longevity of the Welsh Basin as a geological entity.

Ordovician tectonics and volcanism
A worldwide reduction in volcanism due to a reduction in plate movement in the early Ordovician led to a global cooling to more normal levels after the 'hothouse' of the late Cambrian. Sea floor spreading had widened the Iapetus Ocean, which lay to the northwest of the present Herefordshire, to a major ocean close to 5,500km wide (Fig. 3.9). The onset of the Ordovician saw the development of major fault-bounded basins, such as the forerunner of the Worcester Basin (Graben), on the north-west margin of Gondwana. This first indication of the formation of a new spreading centre initiated the breaking away of the Avalonian micro-continent from its parent supercontinent. These marginal basins were themselves an integral part of the initial formation of the proto-Rheic Ocean between Avalonia and Gondwana. The opening of this proto-Rheic Ocean moved Avalonia northwards, with subduction initiated on its northern side, leading to the first active island arc of the Iapetus Ocean in late Tremadoc times.

Ordovician strata and intrusions in Herefordshire
It was in these deepening basins that large thicknesses of fine-grained sediments accumulated. In Herefordshire the Bronsil shales succeed the Whiteleaved Oak Shale with no apparent discontinuity. Because the local dip is to the west, they lie in an area to the west of the Cambrian rocks.

They contain the earliest dendroid graptolite *Rhabdinopora flabelliformis* of Lower Ordovician age as well as trilobites and brachiopods. Graptolites were floating organisms made up of branches along which were arranged minute cups in which the individuals within the colony lived. They formed an important part of the plankton in both Ordovician and Silurian seas. Their rapid evolution and widespread distribution has led to their extensive use in dividing these Palaeozoic rock sequences into the finest time divisions, called zones, and their use as zone fossils (with each zone named after its particular graptolite species).

Deposited under less reducing conditions than the Whiteleaved Oak Shales, the Bronsil Shales contain less carbonaceous matter of organic origin and so tend to be of a lighter, silver-grey colour. They are penetrated by a series of igneous intrusions of both doleritic and andesitic composition.[5] These show contact metamorphic effects against both sets of shales (the shales are baked to a white colour by the heat of the nearby intrusion) and are not seen to pass into the overlying Silurian strata so are assumed to be of Ordovician age. They represent the only indication in our county of the voluminous Upper Ordovician volcanism of

Snowdonia and the Lake District generated by the overall south-easterly directed subduction zone of the closing Iapetus Ocean to the north-west. The intrusions form distinct ridges and isolated bosses within the valley south of Obelisk Hill, for example near Fowlet Farm at SO 7550 3623. They are especially clear during hot, dry summers when their surfaces, lacking a deep cover of soil, are the first to show desiccated grassland.

Three main types of intrusion have been identified[6] all containing sodium-based plagioclase feldspar, which indicates a relationship to submarine basalt lavas (spilites). Minor contact metamorphism in the form of pale-coloured bands of hornfels (see Box 3.3) is seen

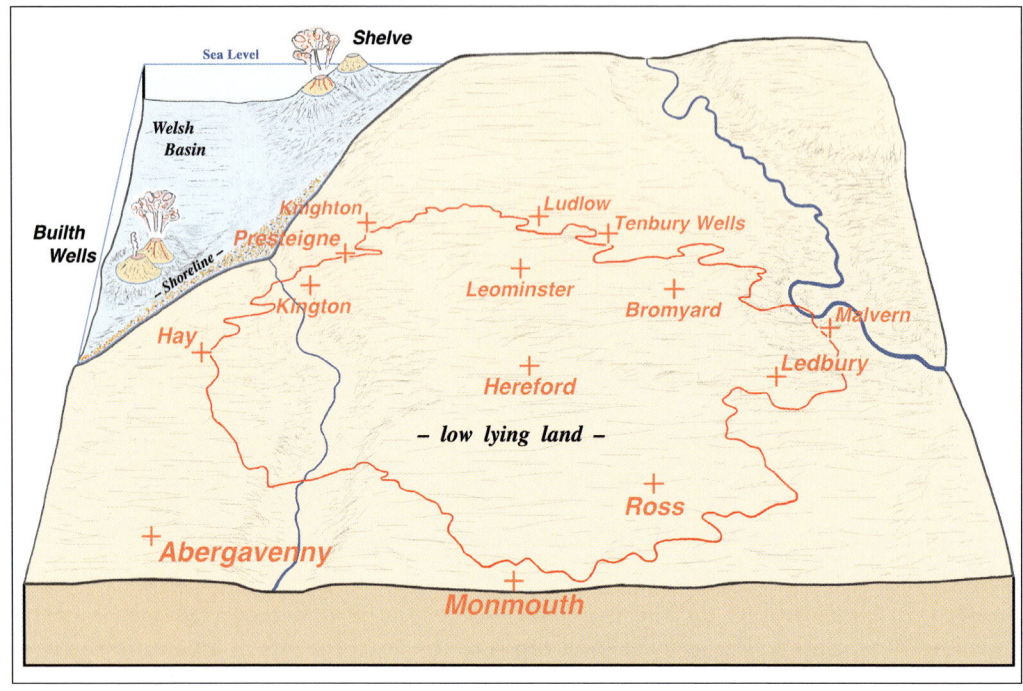

*Figure 5.7 Diorama of Herefordshire palaeogeography during middle Ordovician times (about 460Ma). The retreat of the sea, underway since the high point of Cambrian times, was complete. The shoreline lay roughly along the line of the Church Stretton Fault, dropping away sharply into the deep waters of the Welsh basin to the north-west, studded with active and inactive volcanic islands, particularly in the neighbourhoods of modern-day Shelve, Builth-Llandrindod Wells, Llanwrtyd Wells and Llandeilo. Explosive eruptions must have showered the nearby Herefordshire coast with ash, but all was probably rapidly eroded. As a small section of the slice of continental crust (Avalonia) that rifted away from Gondwana in lower Ordovician times, Herefordshire was part of a low, probably undulating, and long-lived (about 30My) landmass. It was possibly supplied with rivers and sediment by an area of higher relief to the south (termed 'Pretannia') that was much later – at the end of Carboniferous times – displaced westwards, and replaced by what is now Devon and Cornwall. The position of the rivers on the diagram is entirely conjectural.
(Drawn by Gerry Calderbank from a conceptual sketch by Dave Green)*

surrounding these intrusions into both Cambrian and Ordovician mudrocks. This is best seen at Coalhill Cottage (SO 756 357) close to Whiteleaved Oak.

A further outcrop of these intrusions occurs on the A438 at Hollybush (SO 757 368), where a spilitic andesite dyke intrudes the Hollybush Sandstone. A dolerite intrusion, displaying prominent onion-skin weathering, is seen on the south side of the road (SO 750 366), near the entrance to Bronsil Castle. (Onion-skin weathering is the result of the flaking off of successive outer layers of exposed rocks. This eventually produces a rounded surface.)

In north-west Herefordshire, soft, yellowish-grey shales and micaceous siltstones of the same age as the Bronsil Shales outcrop east of the Precambrian Brampton Formation, near Pedwardine (SO 365 711).[7] The graptolite *Rhabdinopora flabelliformis* has been identified, confirming their Ordovician age. These rocks are seen to be overthrust from the west by Precambrian (Longmyndian) rocks.

Local deformation of the coastline of the Welsh Basin occurred as it moved north at the end of the Tremadoc. This is thought to have been caused by an oblique collision of the subducting Iapetus ocean crust beneath the marginal basins (grabens) and their intervening horst blocks. The resultant uplift led to a retreat of the sea and thus Herefordshire became an area of erosion rather than sediment deposition (Fig. 5.7).

For most of the remainder of the Ordovician, the shoreline of the Iapetus Ocean lay to the west of the Pontesford-Linley Fault (an element of the Church Stretton Fault System) resulting in no deposition in the Welsh Borders. Active volcanism was associated with subduction on the edge of Avalonia, now represented by the Lake District, while the Welsh Basin's activity (in north Wales and Pembrokeshire) owes its origin to back-arc volcanism, where the subduction zone was generating extension and rifting. The late Ordovician saw the only marine incursion to the east of the Pontesford-Linley Fault but deposition is restricted to south Shropshire and no inundation affected Herefordshire.

The microcontinent of eastern Avalonia continued to move northwards throughout the Ordovician and eventually collided with Baltica, a larger continental fragment, at the end of Ordovician times. This 'Shelveian event'[8] caused substantial folding and faulting in the Welsh Borders and a major uplift of the margins of the Welsh Basin. It appears to have induced some early thrusting on the Malvern fault system, raising the rocks of the present Malvern Hills by perhaps some hundreds of feet.[9] The sea retreated not only due to this local uplift but also the onset of the end-Ordovician glaciation centred on what is now Morocco in North Africa, at that time located near the South Pole. This caused a worldwide fall in sea level as the glaciers advanced.

By the end of the Ordovician Period the southern half of Britain, as an integral part of eastern Avalonia, had joined with Baltica and moved to a latitude close to 30°S. The Iapetus Ocean, between Avalonia and Laurentia, was still closing and was by that time only 1,000km wide. With the melting of the Saharan Ice Cap, a worldwide marine inundation heralded the succeeding Silurian Period and its very distinctive contribution to Herefordshire's geology.

Further reading

The Rural Geology & Landscape Trail guide to Chase End Hill at the far south of the Malverns covers the Precambrian, Cambrian and Ordovician of this interesting and complex area. It is published by Gloucestershire Geology Trust.

Also see the Geological Conservation Review, vol. 18 (2000) *British Cambrian to Ordovician Stratigraphy*, Rushton, A.W.A. *et al.*

6 The Marine Silurian

Herefordshire occupies a special place in the development of the science of Geology, for this is where Roderick Impey Murchison embarked on tackling what in his day were the frontiers of the discipline of stratigraphy – the study of the stratified, or bedded, rocks. The rocks below the Old Red Sandstone were generally much distorted, metamorphosed and generally lacking in fossils, and had been broadly designated as 'Primary', 'Ancient' or 'Greywackes'. They were thought to be too difficult and intractable to treat in the same way as younger series. Murchison (with Adam Sedgwick) set out to bring intellectual order to the upper parts of this sequence, the so-called 'Transition Rocks'. As he wrote, retrospectively, in his volume '*Siluria*' (1854):

> Desirous of throwing light on this dark subject, I consulted my valued friend and instructor, Dr. Buckland, as to the region most likely to afford evidence of order, and by his advice I first explored, in 1831, the banks of the Wye between Hay and Builth. Discovering a considerable tract in Hereford, Radnor and Shropshire, wherein large masses of grey-coloured strata rise out from beneath the Old Red Sandstone, and contain fossils differing from any which were known in the superior deposits, I began to classify these rocks. After four years of consecutive labour, I assigned to them (1835) the name Silurian.

Murchison derived the name 'Silurian' from the Romano-British people who used to inhabit the region to the south. Murchison was a great supporter of the pioneer geologist from the previous generation, William Smith, who was the first English geologist to recognise that the relative age of beds of sediment, even those of different types (like sandstone and limestone) could be ascertained by the assemblage of fossils found in them. The great advantage of the rocks found in Herefordshire and neighbouring counties was that they were relatively undeformed and contained an abundance of fossils (Figs. 6.1 and 6.2), which Murchison used to construct a sequence that could be recognised across the region and, eventually, much further afield.[1]

This area continues to be important in Silurian geological science. As will be described in this chapter, Herefordshire contains two sites of worldwide significance and most of the names of the major divisions of the Silurian Period are derived from local place names.

At the beginning of Silurian times, 444 million years ago, the area of the Earth's surface that we now call Herefordshire would not have been recognisable to us. The district lay about 20° to 30° south of the equator, therefore in sub-tropical latitudes, roughly where we

Figure 6.1 (above) Representative Silurian fossils : Brachiopods. (© Dave Green)

Figure 6.2 (below) Representative Silurian fossils: Corals, trilobites, crinoids, nautiloids, stromatoporoids, bryozoans and graptolites. (© Dave Green)

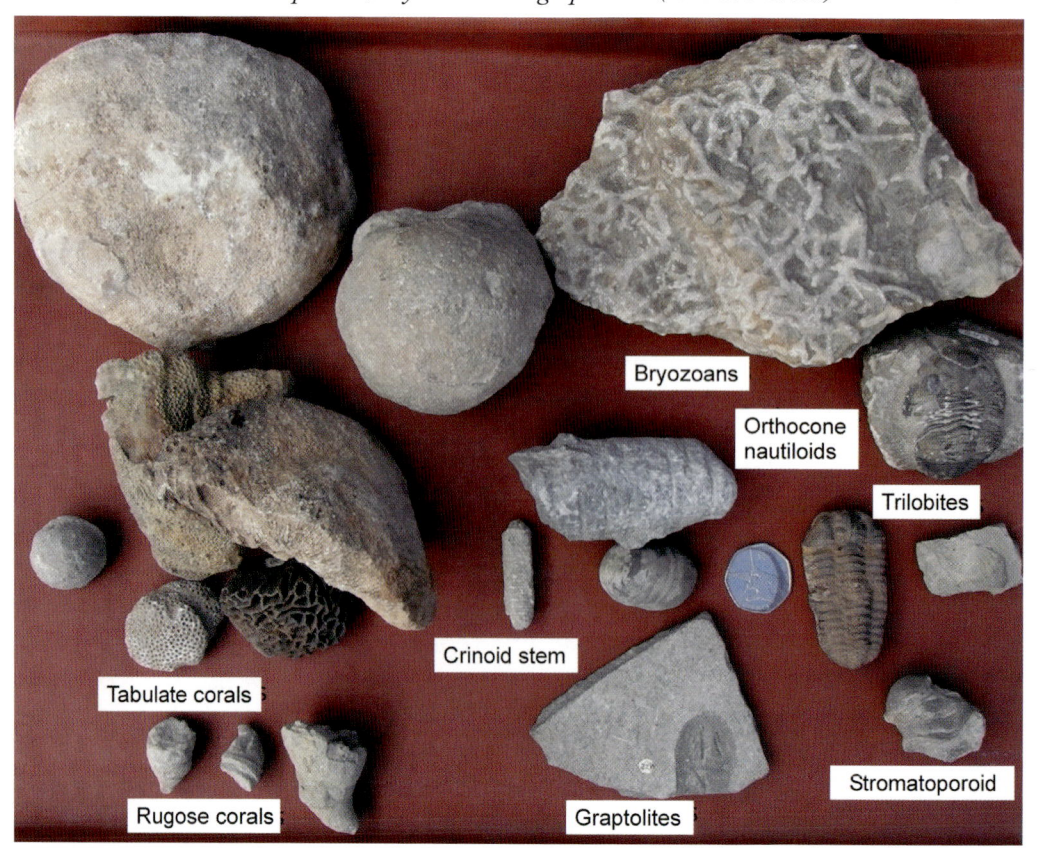

would find Capetown, Perth or Brisbane today. To the north lay the narrow remnants of the Iapetus Ocean (or 'proto-Atlantic', Iapetus being the father of Atlas, the king of Atlantis in Platonic mythology). This previously large body of water dividing us from Laurentia (which contained what are now North America and Scotland, the latter lying off the southeast coast of what is now Greenland) was now shrunk to a tiny remnant, perhaps 100km or so across, by the consumption of its floor down ocean trenches (Fig. 3.10). One of these was still active to the north-west, bringing us ever nearer to Laurentia. To the north-east lay Baltica, a continent comprising what is now Scandinavia, together with Russia to the west of the Urals. Since the late Ordovician, Baltica was attached to the 'microcontinent' of Avalonia, including what is now England and Wales, Southern Ireland, the Low Countries and Northern France, together with southern Newfoundland and parts of Nova Scotia. To the south lay the still widening Rheic Ocean, initiated by the rifting away of Avalonia from the supercontinent of Gondwana (Fig. 3.10). There may well have been a land area, dubbed 'Pretannia', where there is now the Bristol Channel and Devon and Cornwall, for most of the Lower Palaeozoic, supplying most of the material making up the sediments of Silurian age.

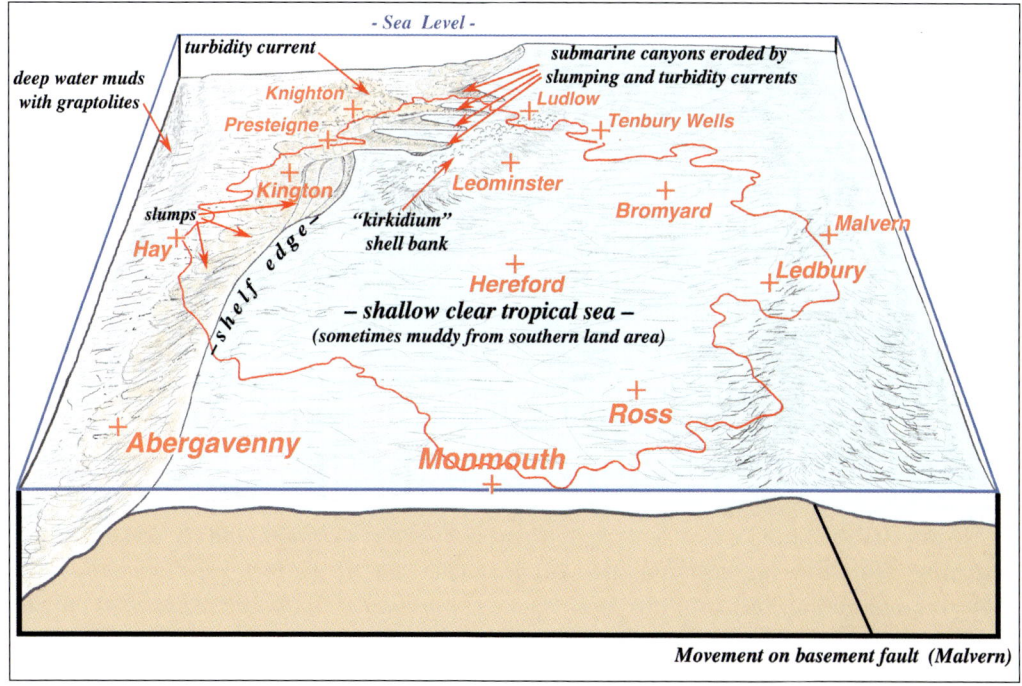

Figure 6.3 A generalised picture of Herefordshire in the Silurian Period. This diagram represents a snapshot in geological time – in this case 425Ma, during the deposition of the Ludlow Series of rocks. The rest of the Silurian Period (which lasted from 444 to 419Ma) would have seen significant differences – the advance of the shallow sea from the west across a low lying land area during the Upper Llandovery (from about 435Ma), and a retreat towards the north and west during the Přídolí (from about 423Ma), exposing, beyond the sandy coastline, a low-lying muddy coastal plain to the south and east.
(Drawn by Gerry Calderbank from a conceptual sketch by Dave Green)

There had just been a glaciation (in the late Ordovician). The ice cap covered the part of Gondwana lying on the South Pole – today forming North Africa! Global sea levels were still low (perhaps by 100m) as a result of the build-up of ice, and that part of Avalonia that is now Herefordshire was above sea level, but only just. It was a very low-lying land area (Fig. 5.7), composed of Precambrian volcanic rocks which were exposed to the east and were overlain by Cambrian and Ordovician sediments to the south and, especially, the west. The nearby coast to the west ran through what is now Welshpool, Rhayader, Llandovery and down to Haverfordwest. This coast marked the edge of a persistently subsiding area of the earth's crust, known to geologists as the Welsh Basin. Land extended as far as the present Northern Pennines to the north, Bedfordshire to the east, and well to the south of Devon and Cornwall. We know this by applying the geological Principle of Uniformitarianism, discussed in Chapter 3, which emphasises the constant nature of geological processes over

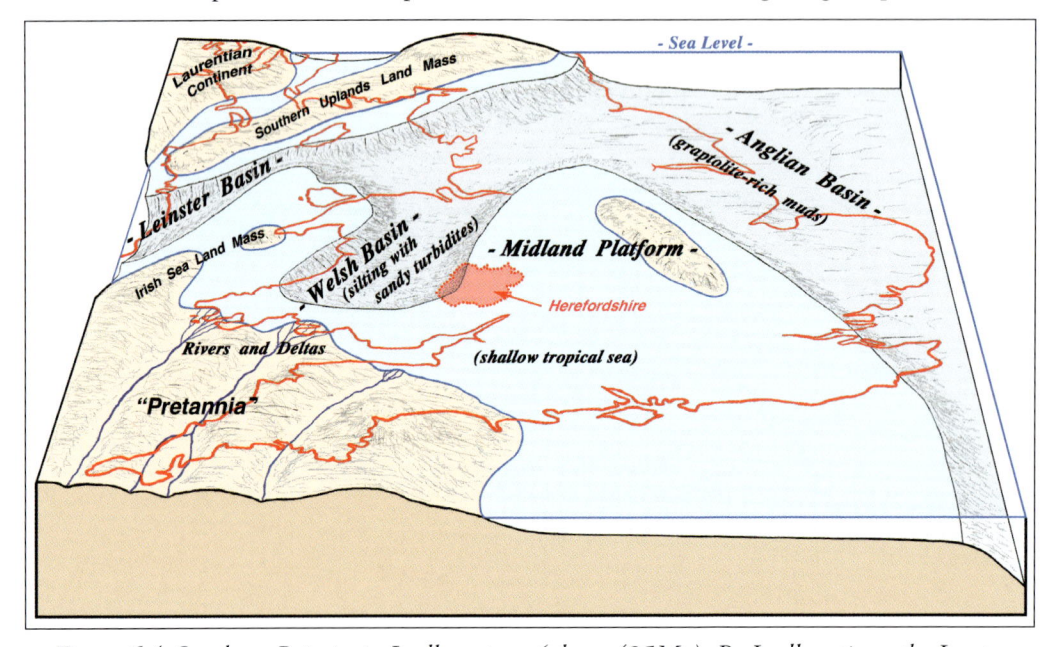

Figure 6.4 *Southern Britain in Ludlow times (about 425Ma). By Ludlow times the Iapetus Ocean had completely closed between northern and southern Britain, and the remaining subsiding basins were rapidly filling with sediment. Note that the collision with Laurentia had been very soft; the only hilly area was probably the pile of slices of scraped-off sediment at the site of the ocean trench – the Southern Uplands of Scotland. However, it is far from clear whether what is now the Southern Uplands was actually where it is today, or whether it was still some distance to the north-east. The collision had not been head on but oblique, causing leftward (sinistral) wrench fault movement along the line of the Iapetus suture. Equally it is unlikely that the landmass to the south of Britain, whose rivers carried the first Old Red Sandstone deposits to Pembrokeshire, was what is now Devon and Cornwall, which were probably sited some 400 km to the south-east (south of where Paris is now!). A major fault – the Bristol Channel-Bray Fault, moved what was the Pretannia landmass westwards so that its remains probably form part of the submerged continental shelf south of Ireland.*
(Drawn by Gerry Calderbank from a conceptual sketch by Dave Green)

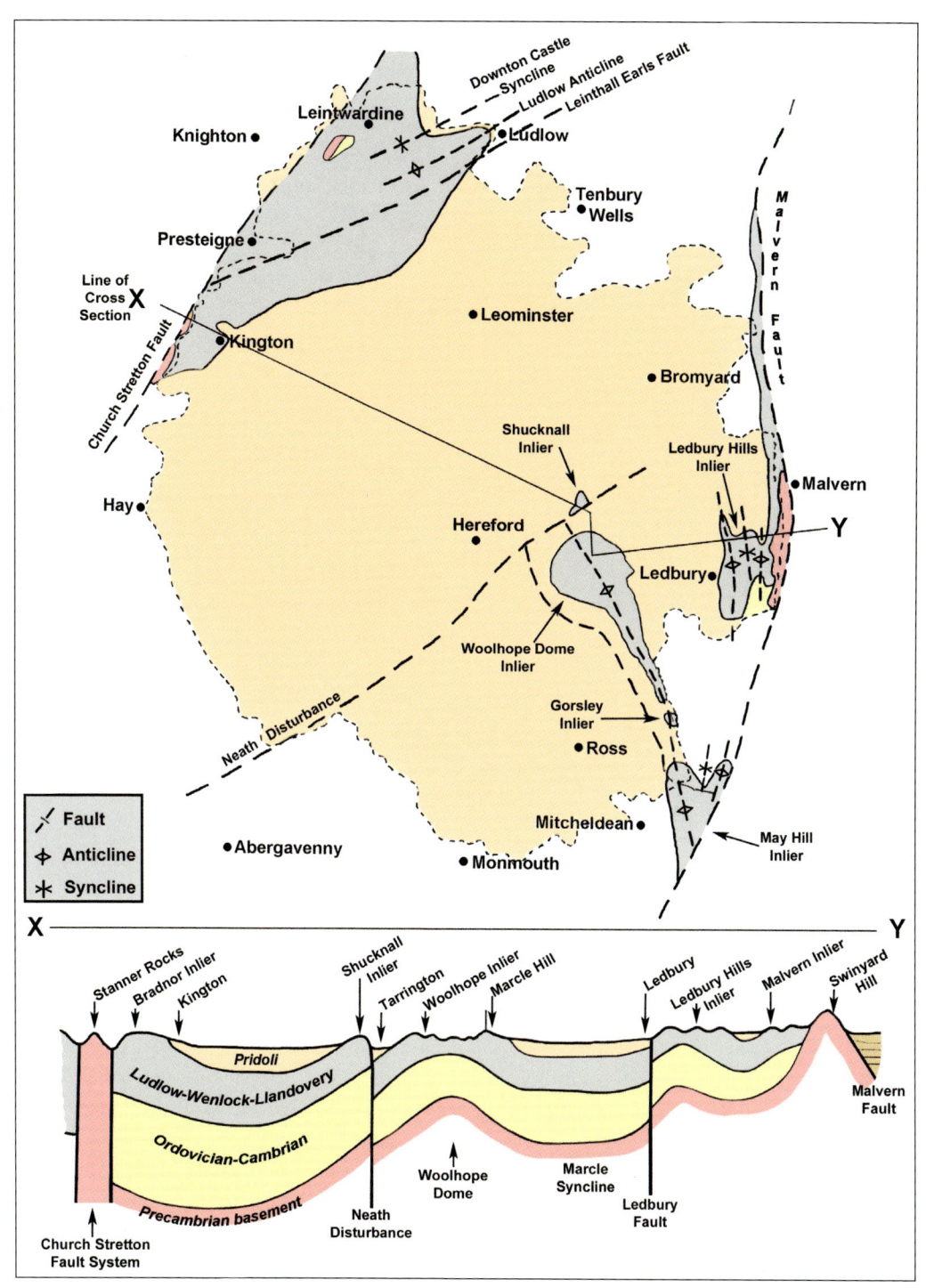

*Figure 6.5 Sketch map and cross section to illustrate the relationships between outcrops of Silurian rocks and underlying geological structure.
(Drawn by Gerry Calderbank from a conceptual sketch by Dave Green)*

Figure 6.6 The view to the north-west over the Woolhope area reveals the underlying dome structure with concentric 'circles' of limestone outcrops marked by wooded areas. The limestone stands up as ridges separated by more easily eroded clay vales. The slope in the foreground is on the dip slope of the highest limestone layer, the Aymestry Limestone. The instability of such areas, with massive limestone resting on a slope of weak shale, leads to occasional landslips. The slope shown here was the site of such a landslide, known as 'the Wonder', in 1571. (© Derek Foxton Archive)

time. Thus, the lack of any deposits in Herefordshire from the earliest Silurian times (and, for that matter, from the late Ordovician) indicates that the area was above sea level during that period, just as for a similar situation today. Thus, either there was no deposition at this period or any earlier marine-deposited rocks were eroded away

A glance at the geological map and cross section of Herefordshire (Fig. 6.5) indicates that the marine rocks of Silurian age outcrop in a number of scattered areas, especially around the periphery of the county, and are separated by younger rocks, mainly the Old Red Sandstone. These isolated outcrops are termed *inliers* and they form many of the hills on the north and east sides of the county, as well as centrally in the Woolhope area. They represent the upper parts of layers of older rock poking through, and surrounded by, the cover of younger rocks. In other words, over the majority of the county, marine Silurian rocks are present, but buried beneath layers of rock laid down on top;[2] in terms of total area the majority of this is the Old Red Sandstone. The existence of the inliers is largely due to upfolding (anticlines and domes), as in the Woolhope Dome south-east of Hereford[3] (See Fig 6.6) and its extension south-eastwards to the Gorsley Anticline.[4] Similar structures are in the May Hill Anticline east of Ross-on-Wye[5] and the Ludlow Anticline (Wigmore Dome) north-west of Leominster.[6]

Most of these folds have been produced by compression due to movement on major faults that have also raised the Silurian rocks to higher levels, where they have been exposed by the erosion of the overlying rock. (Both the folds and the faults are results of the Variscan orogeny in the Carboniferous Period and are discussed in Chapter 8.) The effects of these movements are shown, for instance, to the west and north of the Malverns in the uplifted and tightly folded western side of the great north-south trending Malvern Fault (Fig. 8.14), and now forming the Ledbury Hills, the Suckley Hills and Ankerdine Hill.[7] East of Hereford, a major north-east to south-west trending fault (the Neath Disturbance) brings Silurian rocks to the surface to the north of the Woolhope Dome as Shucknall Hill. In the north-west of the county, movement on the equally important Church Stretton Fault has resulted in the dragging up of the Silurian rocks making up Hergest Ridge and Bradnor Hill, but on the downthrown side of this fault (Fig. 6.5).

As the late Ordovician ice melted and the Welsh Basin continued to subside, the sea began to encroach on the low land area from the west, laying down first beach deposits. These are usually sandstones and conglomerates, for instance the Folly Sandstone near Presteigne, the Huntley Hill Formation in the May Hill area, and the Cowleigh Park Formation in the Malverns. These sometimes contain shelly fossils, especially brachiopods of the '*Orthis*' type, and the very long-lived *Lingula* – still living today in burrows under the beaches and inshore sands of Japan!

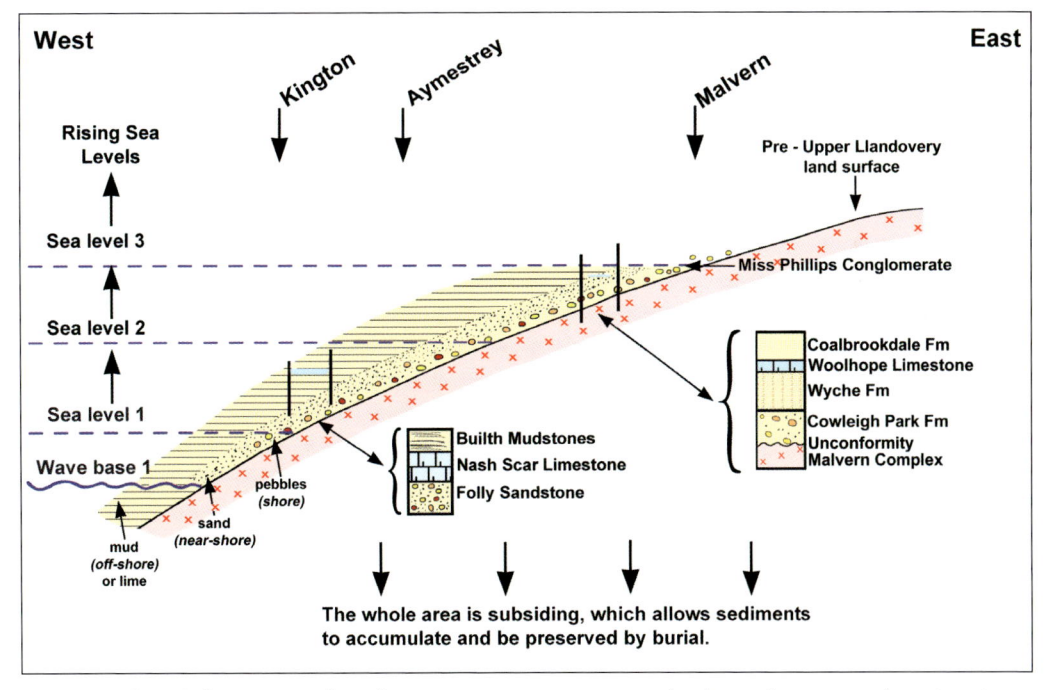

Figure 6.7 A diagram to show how a marine transgression leads, with time, to the migration of beach, nearshore and offshore environments inland, and hence a distinctive fining upwards sequence of sediments, illustrated by the Lower Silurian (Upper Llandovery) Transgression. (Drawn by Gerry Calderbank from a conceptual sketch by Dave Green)

Since changes of sea level (due to changes in the volume of ice on land or the capacity of the ocean basins) are the rule rather than the exception over periods as long as geological time, we may expect geological history to be littered with the results of such changes, as land areas were alternately drowned and exposed. A general description of these marine transgressions and regressions is given in Box 6.1, where the use of fossils in tracing their occurrence is noted. In a ground-breaking study researched in the 1960s, in Herefordshire, Shropshire and Gloucestershire, Ziegler, Cocks and McKerrow[8] identified marine communities of the Lower Silurian (Llandovery), and, by plotting their data on maps for each time division, showed exactly where and when the sea reached certain depths as its shore line swept slowly eastwards across the old land surface. Further studies were carried out for the middle (Wenlock) and upper (Ludlow) divisions of the Silurian.

Stratigraphical terminology

The naming of Silurian strata in Herefordshire has varied considerably over the years, particularly the last 60 or so. In this book the most recent, and thankfully simplest, version of this is used. If other texts are consulted different names will be encountered for the same rocks. This section gives an account of the variety of the principal names and shows their correlations. The section may be easily omitted by the more casual reader.

Box 6.1   **Marine Transgressions and Regressions**

Drowning of the land by the sea is known as a **marine transgression** and in the early Silurian the process took quite a long time (up to 10 million years!) and included periods when sea level fell temporarily – minor **marine regressions** – before continuing to rise.

Two important clues can help us to come to this conclusion. Firstly, the type of sediment: as the water deepens, the amount of wave energy affecting the sea floor diminishes. This leads to finer and finer-grained sediments, as can be seen today on modern coasts (uniformitarianism again!). The top of some such beaches is constructed from pebbles thrown up by high-energy storm waves. The lower beach is made of sand, becoming finer still below low tide level, until a depth is reached (wave base) where the water is not moving at all next to the sea floor and mud can settle out of suspension. If we add time to this scenario: as the water deepens over a particular place, what was once beach (e.g. pebbles) becomes near-shore (e.g. sand), and then becomes offshore (e.g. mud), resulting in a vertical sequence of deposits that becomes finer upwards. The opposite would apply to a marine regression (i.e. a coarsening-up sequence). Additional in this type of detective work is the presence of **sedimentary structures**, for example the ripple marks in sand, that everyone will have walked across on a beach, indicate very shallow water oscillating back and forth as waves pass overhead.

Secondly the fossil content of the sediment: most people will be familiar with the idea that as the tide goes out on a rocky coast, it reveals a shoreline that has different flora and fauna near low tide mark, such as large ribbon type kelp seaweeds, sea anemones and sea snails, to that near high tide level, such as bladder wrack and limpets. Similar changes take place to the fauna living on and in the sediments, with different communities of animals specialising in the conditions at different depths.

Over the years since the original naming by Murchison in the middle of the 19th century a plethora of local names has been applied to the divisions of the Silurian, particularly the Ludlow strata. Figure 6.9 shows the correlation between the old and recent naming schemes. Names from the mid-19th century are shown in columns 1 and 2. Much work was done in the 1960s and 1970s[9] to rationalise and extend Murchison's original scheme (Fig. 6.9, col. 7). The bulk of this work was carried out in Mortimer Forest by members of the Ludlow Research Group, an important group of workers in the Silurian rocks.[10] Mortimer Forest lies south-west of Ludlow and mainly in northernmost Herefordshire. Here, careful logging of the sequences of rocks exposed in newly created forestry tracks enabled these workers to identify 'type localities', where they defined the boundaries between the new bundles of beds they identified on the basis of both rock type and contained fossils. These localities may have worldwide significance if accepted by international panels of geologists. They are known as 'Stratotypes' and are identified in a series of published trails that can be followed. In particular, in a disused quarry, near an isolated cottage in the woods known as Pitch Coppice, lies the world stratotype for the boundary between the Ludlow and Wenlock divisions of the Silurian (SO 4724 7301) (Fig. 6.8).[11]

Half a mile further east, towards Ludlow, lies the type locality (Gorsty Knoll) from which one of the divisions of the Ludlow was named 'Gorstian' – which geologists the

Figure 6.8 Pitch Coppice Quarry, the internationally recognised definitive location for the boundary between rocks of the Wenlock series and the Ludlow Series, under inspection by the author of this chapter. (© John Payne)

world over will recognise as a span of geological time. These are important localities and the exercise carried out in the 1960s and '70s was one of the first to apply a rational scheme to stratigraphy that could be used worldwide. Unfortunately, changes in the environments of deposition meant that in other areas corresponding changes in rock type and fossils did not occur, or occurred at different times. This made such a precise scheme difficult to implement widely – hence the variety of different names (Fig. 6.10), often introduced by the same workers when they studied other areas, such as Woolhope or May Hill!

After a period of trying to use names based on the Shropshire sequence (Fig. 6.9, col. 7), particularly on the changes in fossils present, there has been a reversion to Murchison's original scheme, to most geologists' relief! (Fig. 6.9, cols. 5,6,8). Thus, the terminology used in this book is that of Fig. 6.9, columns 4, 5 and 6, although this has not yet (in 2016) been given formal status by the British Geological Survey.

1	2	3	4	5	6	7	8
Systems	Series	Rock Units					
Murchison 1839	Sedgwick 1852	Lapworth 1879	Modern BGS	MODERN BGS	Modern BGS (Formations)	C.H.Holland et al 1950s - 60s	Murchison 1839
OLD RED SANDSTONE	OLD RED SANDSTONE	OLD RED SANDSTONE	ORS	PŘÍDOLÍ	Moor Cliffs (Raglan Mudstone)	Temeside Beds	Cornstone and Marl
					Downton Castle Sandstone	Downton Castle Sandstone	Tilestones
UPPER SILURIAN	SILURIAN	SILURIAN	SILURIAN	LUDLOW	Upper Ludlow Siltstone	Whitcliffe Beds	Upper Ludlow Rock
						Leintwardine Beds	
					Aymestry Limestone	Upper Bringewood Beds	Middle Ludlow Rock
					Lower Ludlow Siltstone	Lower Bringewood Beds	Lower Ludlow Rock
						Elton Beds	
				WENLOCK	Much Wenlock Limestone	Wenlock Limestone	Wenlock Limestone
					Coalbrookdale	Wenlock Shale	Wenlock Shale
					Woolhope Limestone	Woolhope Limestone	
LOWER SILURIAN	UPPER CAMBRIAN			LLANDOVERY	May Hill Sandstone	Various local terms e.g. Huntley Hill Beds, Haugh Wood Beds	Caradoc Sandstone
	MIDDLE CAMBRIAN	ORDOVICIAN	ORDOVICIAN	ASHGILL			Llandeilo Flags
				CARADOC			
				LLANVIRN			
				ARENIG			
CAMBRIAN		CAMBRIAN		TREMADOC			

Figure 6.9 A table to show the development of ideas on the divisions of the Lower Palaeozoic Era. In fact, there were slight discrepancies between the positions of some of the boundaries but, for simplicity, only the major ones have been shown here. Note the beds that were the subject of bitter dispute between Murchison and Sedgwick (columns 1, 2) – a controversy that was only resolved after they had both died … by the introduction of the Ordovician Period by Lapworth (col.3).
(© John Payne)

AGE	THE MARINE SILURIAN					
	GENERAL WELSH BORDERLAND	WOOLHOPE	LEDBURY MALVERN	MAY HILL GORSLEY	LUDLOW	KINGTON
PŘÍDOLÍ	Moor Cliffs fm	Raglan Mudstone Formation				
	Downton Castle Sandstone fm	Rushall fm	Downton Castle Sandstone fm	Cliffords Mesne sst Gorsley sst	Downton Castle Sandstone fm Platyschisma sh	Bradnor sst Platyschisma sh
LUDLOW		Ludlow Bone Bed				
	Upper Ludlow Siltstone fm	Upper Perton	Upper Ludlow Siltstone fm	Upper Longhope	Upper Whitcliffe	(ABEREDW FM)
		Lower Perton		Lower Longhope	Lower Whitcliffe	Llan-wel Hill fm
		Upper Bodenham		Upper Blaisdon	Upper Leintwardine	Wern Quarry fm
		Lower Bodenham		Lower Blaisdon	Lower Leintwardine	
	Aymestry Limestone			gap	Upper Bringewood	Knucklas Castle fm
	Lower Ludlow Siltstone fm	Lower Sleaves Oak	Lower Ludlow Siltstone fm	Upper Flaxley	Lower Bringewood	(CWM GRAIG DDU)
		Upper Wootton		gap	Upper Elton	(IRFON)
		Lower Wootton		Lower Flaxley	Lower Elton	
WENLOCK	Much Wenlock Limestone Formation					(BUILTH MUDSTONE)
	Coalbrookdale Formation					
	Woolhope Limestone Formation				Buildwas Shale fm	Nash Scar lst
UPPER LLANDOVERY	May Hill Sandstone fm	Haugh Wood fm	Wyche fm	Yarleton fm	Hughley Shales	Folly Sandstone
		Huntley Hill fm	Miss Phillips Cgl	Huntley Hill fm	Pentamerus Beds	
		Fownhope fm	Cowleigh Park	– ? –	Kenley Grit	
	UNCONFORMITY					
ROCKS REVEALED IN THE ERODED UPPER LLANDOVERY LAND SURFACE	ORDOVICIAN Caradoc Series CAMBRIAN	CAMBRIAN PRECAMBRIAN Malverns Complex		SILURIAN Huntley Quarry fm Lower Llandovery	CAMBRIAN PRECAMBRIAN Uriconian/Longmyndian	PRECAMBRIAN Old Radnor?

Figure 6.10 A table to show the local names employed in the late 1900s for the various Herefordshire rock units of the Marine Silurian, now taken as extending to the top of the Downton Castle Sandstone Formation. The multitude of local names stemmed from the general inapplicability of the naming system derived for the Ludlow area.
(Notes: 1. Rock units in the Kington area have not yet been assigned formal names. Those in brackets are from the adjoining areas in Wales.
2. In the Ludlow area, the Temeside Mudstone Formation lies between the Downton Castle Sandstone and the Moor Cliffs Formation.
3. Cells of identical colour in Figs. 6.9 and 6.10 have identical meanings.
4. The General Welsh Borderland column in yellow is now the standard for the region, though not yet formally adopted by the British Geological Survey and so liable to further changes. Many geologists still use the older, local names in a generally unsystematic manner.)
(Drawn by Gerry Calderbank from a conceptual sketch by Dave Green)

The Silurian rocks of Herefordshire

The Llandovery Series
The earliest rocks of the Silurian exposed in Herefordshire are of the **Upper Llandovery** sands laid down at various times during the marine transgression. These sands are given the collective name of **May Hill Sandstone Group**, after their type locality. Due to erosion of the underlying land surface, these sands lie on older rock of various ages, from Precambrian, as in the exposure high up in the north-west corner of Gullet Quarry (SO 7613 3822) on the southern end of Swinyard Hill, through Cambrian and Lower Ordovician near Bronsil Castle (SO 749 372), east of Ledbury. In most cases these rocks form high ground,

indicating their resistance to erosion, as on May Hill (SO 696 213), Rough Hill Wood (SO 759 483), the high ridge near Eastnor on which stands Lord Somers' obelisk (SO 752 378), and The Beck (SO 749 505) north of Great Malvern. They are surprisingly poorly exposed. The cement holding the grains together rapidly breaks down near the surface and they have been little used as building materials. A small quarry (SO 697 222) on the northern slopes of May Hill above Clifford's Mesne exposes yellowish-weathering sandstone of the Huntley Hill Formation with beds of gravelly conglomerate. At the top of May Hill (just in Gloucestershire), some of the reputed one hundred trees have been uprooted, exposing sandstone and conglomerate debris, sometimes with brachiopods in the finer sandstones. The Cowleigh Park Formation of the Malvern area is usually greenish grey quartz sandstones and conglomerates, exposed scrappily near the road at SO 760 479. In its upper part, for instance at the football pitch in High Wood, West Malvern (SO 7620 4738) it consists of red, gritty sandstone with conglomerates; the red colour indicates a temporary regression so that the area became land before re-submergence. In the north-west, where the sea arrived earlier, the high ground of Nash Wood (SO 304 627), south-west of

Figure 6.11 Nash Scar Quarry, between Kington and Presteigne, viewed from the south-west. The left-hand side of Nash Scar quarry shows almost horizontally bedded Folly Sandstone, with the overlying Nash Scar Limestone (equivalent to Dolyhir or Woolhope Limestone) mainly removed by quarrying, apart from low faces just visible below the conifers. A minor branch of the Church Stretton Fault system runs away from the camera, faulting Nash Scar Limestone on the right down against the Folly Sandstone. The movement has caused the ends of the limestone beds to be dragged up next to the fault line, explaining the relatively steep dip to the right (south-east). The quarry was worked over a very long period: Murchison mentions it as a 'magnificent' natural rock outcrop in the 1830s, partly quarried for limestone. It supplied a large area of limeless central Wales and provided building stone to the local area – Presteigne Jail had then been recently constructed of Folly Sandstone. (© Dave Green)

Presteigne, is formed from the Folly Sandstone of older age (lower to middle Llandovery). This is a dark grey often pebbly sandstone, well exposed beneath the middle Silurian Nash Scar Limestone, at the south-western end of Nash Scar Quarry (SO 301 623) (Fig. 6.11).

The sediments deposited on top of the sands reflect a deepening of the water as the sea invaded further east. So, on May Hill, the Huntley Hill Formation is overlain by the finer-grained sands, silts, muds and thin limestones of the Yartleton Formation, the equivalent in the Malvern area being the Wyche Formation and in Woolhope, the Haugh Wood Formation. The relatively calm, undisturbed shallow water allowed a more diverse fauna to thrive. Fossils are frequently found, especially brachiopods, crinoids and solitary 'cup' corals. Undoubtedly one of the best places to see these beds is in the top north-west corner of Gullet Quarry (SO 7613 3822), a huge aggregate quarry worked mainly in the 1960s and '70s for building the M5 and M50 motorways. If you climb to the top of the path here, past an area (on your right) heavily fenced to try to stop the local youths jumping to their potential death in the icy water below, you will reach this small quarry. Here, the bottom beds of the Wyche Formation can be seen to be lying almost vertically. There is (in 2015) a prominent but decaying thick bed of sandstone in the middle of the sequence, with clearly visible wave ripple marks on its upper (west) side, indicating that the water must have been shallow. Brachiopod shells and horn-shaped cup corals, also indicating shallow water, can be found in the abundant small blocks that have been weathered from the face or, at least in most cases, the mould of their shape left in the rock – modern weathering by slightly acid rainwater has dissolved the calcium carbonate of which they were once composed. If you look carefully to the right, you will see the contact between the Silurian rocks and the Precambrian diorites, which is irregular and contains rounded boulders and pebbles of diorite and granite, presumably formed on the beach at the time. This is another unconformity, which is also exposed further north, most famously at the 'Sycamore Tree Quarry' (SO 7647 4594) where the 'Miss Phillips Conglomerate' rests on Malvern diorite. The Miss (Anne) Phillips in question was the geologically astute but much less well known sister of John Phillips, Professor of Geology at London and Dublin. He surveyed the Malverns in the early 1840s for the Geological Survey and was William Smith's nephew. Anne's discovery[12] disproved the earlier interpretation (by R.I. Murchison) that the Malvern igneous rocks had been forced up as liquid magma into the Silurian rocks; they had obviously been cold and exposed as rocks being eroded by waves on the advancing shoreline prior to the Silurian deposition.

The Wenlock Series
At the beginning of **Wenlock** times (433 million years ago), the sea continued to deepen and the area of the old land surface to shrink towards the east, away from Herefordshire. The county was now at the bottom of a relatively shallow, well-lit, clear tropical sea whose waves hardly stirred the fine-grained sediments accumulating on its floor. With little land nearby to produce sand or clay, the majority of this sediment came from calcium carbonate (lime) dissolved in the water and precipitated from it in these warm, still conditions. This is known as the **Woolhope Limestone** at Woolhope, May Hill and the western side of the Malverns. It consists of olive grey muddy limestones, siltstones and four thin yellow clays known as bentonites, which are the fine-weathered remains of the ash fall from distant

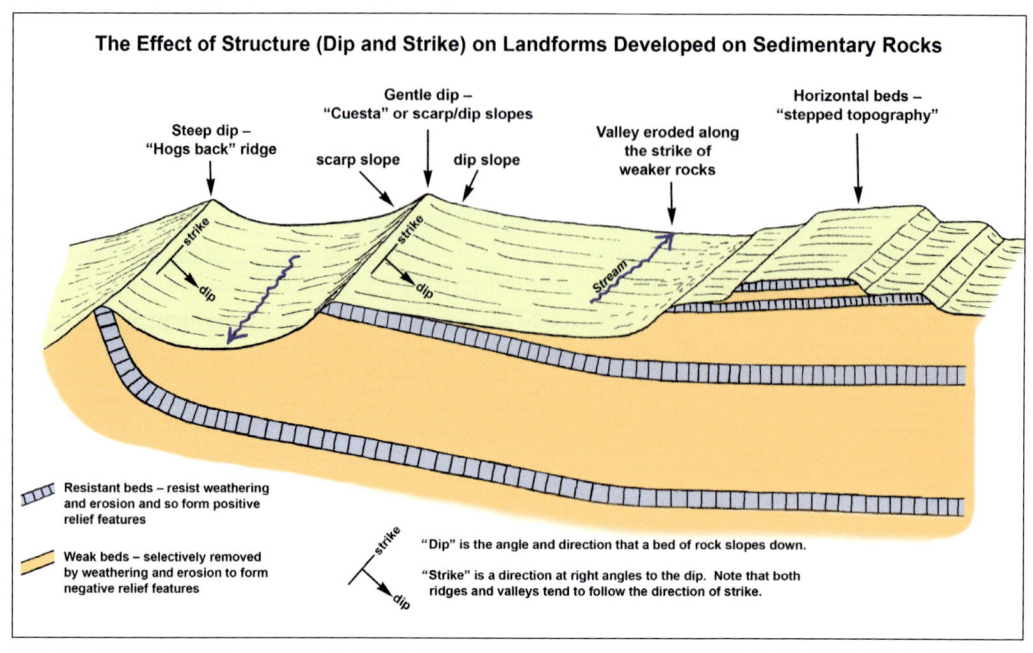

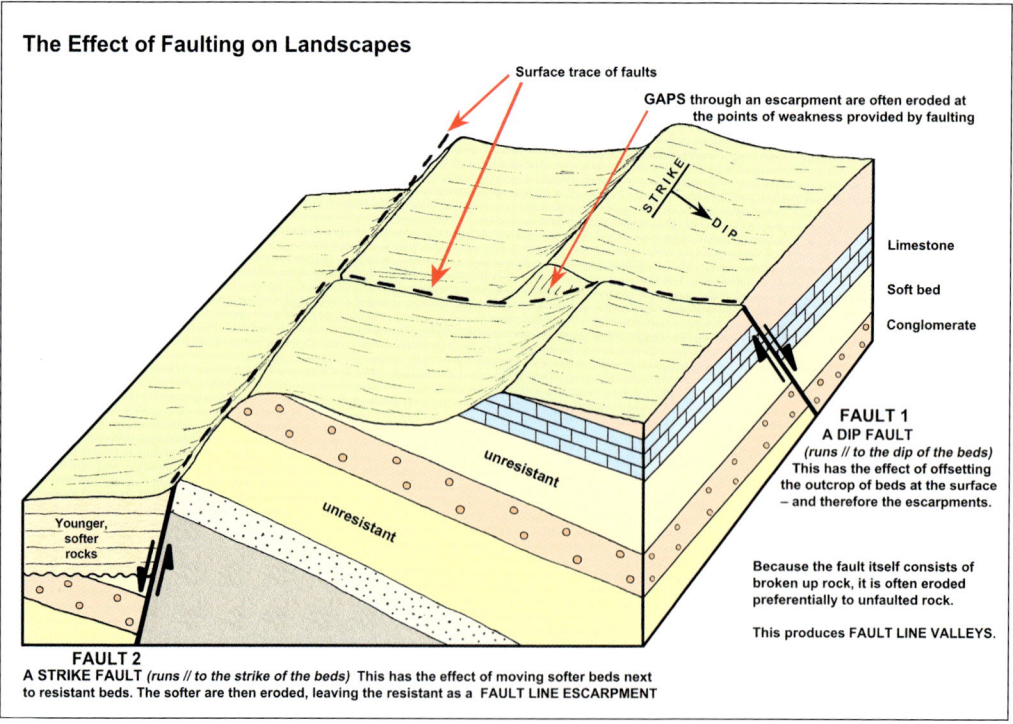

Figure 6.12 Diagrams to illustrate the relationships between geological structure and landscape: a) the effect of dip and strike, b) the effect of faulting. (Drawn by Gerry Calderbank from a conceptual sketch by Dave Green)

volcanoes. Good exposures of this limestone are found at the type locality, an old quarry near the village of Woolhope (SO 612 357), and Scutterdine Quarry (SO 5778 3854), which is even better, and very important since it forms one of the stratotypes or reference sections from which this particular rock unit is defined! Knight's Quarry (SO 754 484) near Storridge provides a good exposure in the Malvern area. Near Presteigne, the large, now abandoned, Nash Scar Quarry (Fig. 6.11) was mainly worked for limestone (Nash Scar Limestone, similar to the nearby Dolyhir Limestone of the same age). The limestone is extensively recrystallised, presumably due to hot fluids rising up the shattered rocks of the fault zone. Strangely, for the most part the fauna of these rocks is very limited in both numbers and diversity, being dominated by tabulate coral colonies such as *Favosites*, *Heliolites* and *Halysites*, solitary corals (especially *Schlotheimophyllum*) and brachiopods of only a few species (*Coolina*, *Dalejina*, *Dicoelosia*, *Eoplectodonta*, *Eocoelia* and *Leptaena*). These fossils are often broken up – suggesting that much of the debris was washed in from shallower areas during storms. Because both the underlying and overlying formations are dominated by soft mudstones and siltstones, the Woolhope Limestone stands out as a ridge (high angle of dip) or escarpment (low angle of dip) in the landscape (Fig. 6.12).

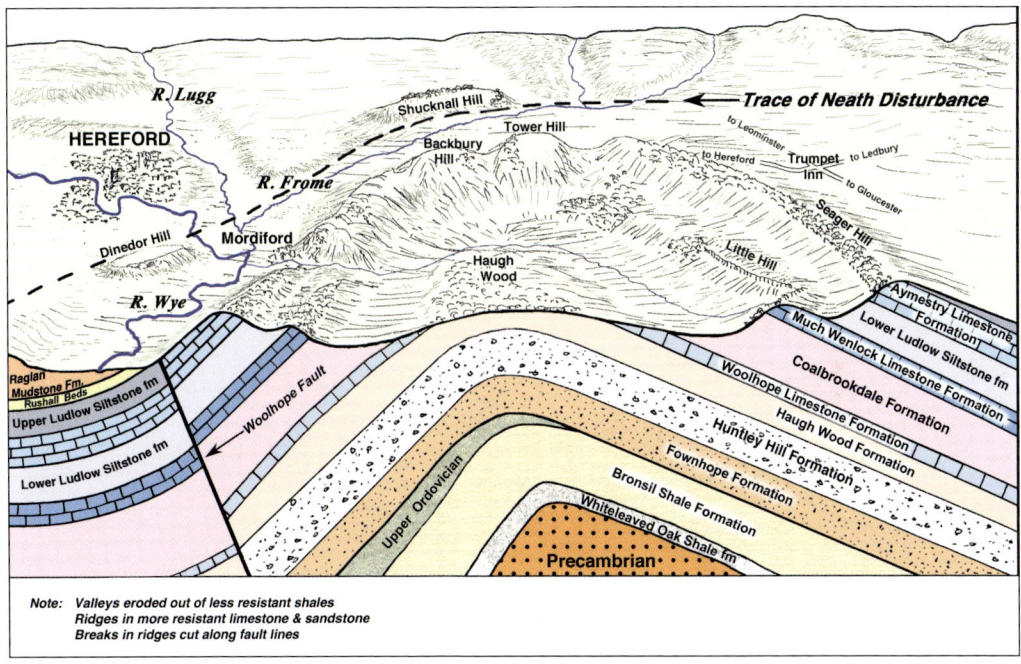

Figure 6.13 Geology and scenery in the Woolhope Dome. This upfold or anticline has been formed by compression due to sticking on the Woolhope Fault. It shows a classic relationship between the underlying geology and the landscape of erosion developed on the surface. It is difficult to find a place from which to see the whole of this wonderful feature. The top of Seager Hill or Marcle Hill in the east, being the highest points would theoretically be best, but the view is too often obscured by trees. It can be viewed at a distance from Perrystone Hill near Ross (SO 635 292), and good, though limited views can be had from Tower Hill in the north, and from an unnamed, isolated hill in the west, known to locals as The Knob at SO 604 341.
(Drawn by Gerry Calderbank from a conceptual sketch by Dave Green)

The sketched landscapes (Figs. 6.13, 6.14, 6.15) attempt to show the very close relationship between rock resistance, structure and landscape wherever the shallow-water Silurian rocks are exposed, across the county.

The next layers laid down in the sea are very thick olive-green mudstones and siltstones of the **Coalbrookdale Formation** (formerly Wenlock Shale), indicating very quiet water in which the fine sediments could settle out of suspension. The total thickness is about 250 metres, in contrast to the 25 metres of Woolhope Limestone. Because they are soft, these rocks are very rarely exposed at the surface and form a low-lying vale, sometimes poorly drained and with heavy soils, between the ridges of the Woolhope and Much Wenlock Limestones. Notable locations include the low-lying centre of the Ludlow Anticline (Wigmore Dome) north-east of Wigmore (Fig. 6.15); the low-lying area of Eastnor Park near Ledbury (Fig. 2.3); and the arcuate valley running around the core of the Woolhope Dome, below the village of Woolhope and running north to Checkley (Fig. 6.13). Occasionally there are temporary exposures; the best and most complete was the A4103 road cutting at Storridge (SO 745 485) but much of this is now grass-covered. A nearby good exposure is in Whitman's Hill Quarry (SO 748 483). Fossils are relatively scarce but of great diversity, there being many types of brachiopod as well as crinoids, corals and trilobites.

After about three million years (433 to 430Ma) of mud deposition the water seems to have cleared and the stage was set for the laying down of one of the most celebrated of Welsh Borderland rocks, the **Much Wenlock Limestone** Formation. It is notable for

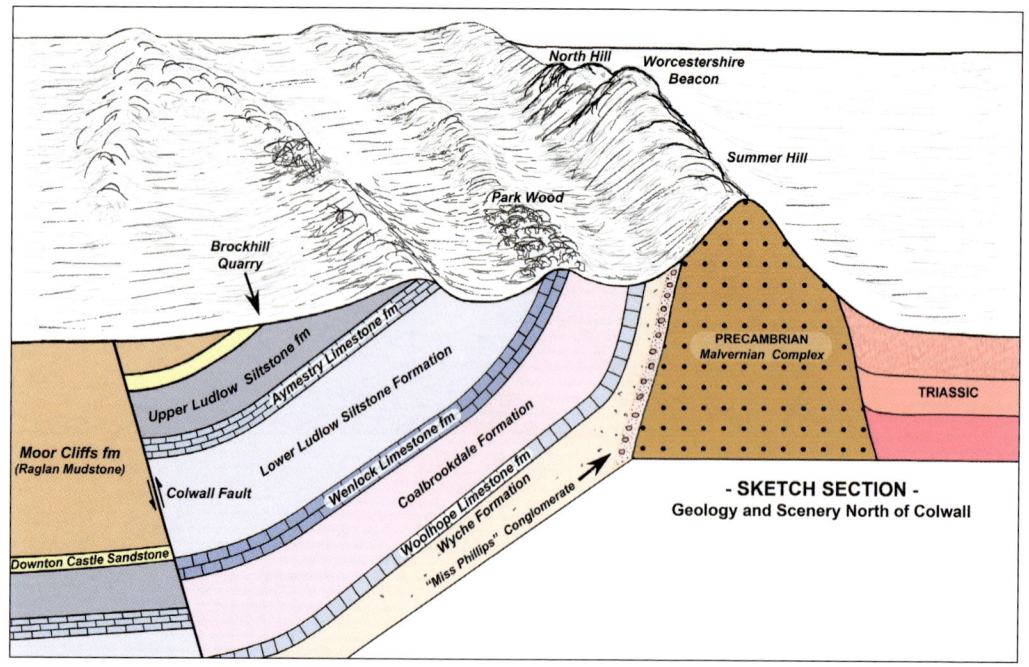

Figure 6.14 Geology and scenery north of Colwall. A sketch to show the development of ridges and valleys on relatively steeply dipping Silurian sediments. An excellent place from which to view this is the top of Herefordshire Beacon (British Camp).
(Drawn by Gerry Calderbank from a conceptual sketch by Dave Green)

a number of reasons. Firstly, it makes a very prominent feature in the landscape (as in the type area of Wenlock Edge, Shropshire). The chief local examples are the inner ring of escarpments in the Ludlow anticline (Wigmore Dome) (Fig. 6.15) and the Woolhope Dome (Fig. 6.13). Also important is the prominent curved wooded ridge of the Ridgeway, west of Herefordshire Beacon, and its continuations south towards Eastnor (Fig. 2.3), north towards Colwall, Storridge and Suckley (Fig. 6.14) and west towards Ledbury (Coneygree Hill). It also makes a prominent ridge feature to the west of May Hill; here it is much broken into segments by faults. Secondly, it is celebrated for the diversity and abundance of its fossil content. In most exposures (of which there are very many, this being a much-quarried rock for lime and stone) you will be able to find a wealth of fossils: corals, stromatoporoids, brachiopods, gastropods, bivalve molluscs, crinoids, trilobites and bryozoans to name the most common. In some exposures, notably Little Hill quarries, north

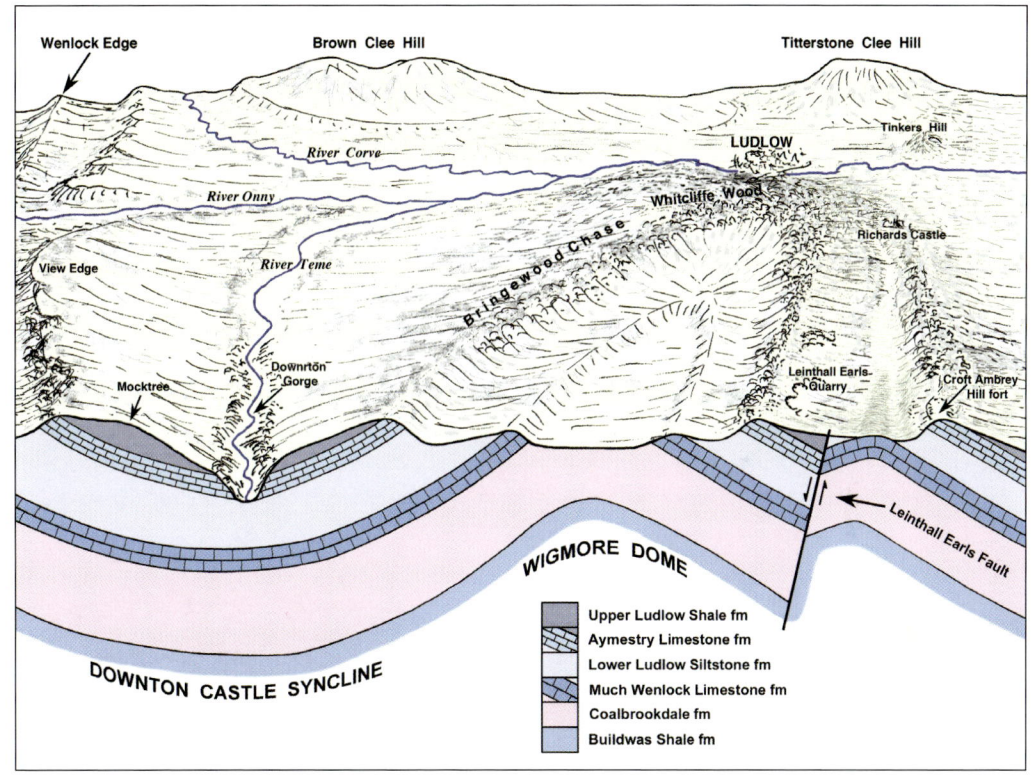

Figure 6.15 Geology and scenery in Downton Gorge and Wigmore Dome (Ludlow Anticline) (looking north-east). In the far north of the county there exists another superb illustration of the relationship between geology and landscape. It is developed on folds associated with the Leinthall Earls Fault, a splay of the Church Stretton Fault system, which here has the effect of repeating the outcrop of the Aymestry Limestone and hence its escarpment. Note that the Downton Gorge is an example of where the scenery is not particularly well adjusted to the underlying geology – the River Teme was diverted from its pre-glacial course to cut right through the resistant Aymestry Limestone (see Chapter 9).
(Drawn by Gerry Calderbank from a conceptual sketch by Dave Green)

Figure 6.16 (above) The view to east-north-east over Croft Ambrey hillfort towards Titterstone Clee Hill in the far distance beyond Ludlow. The ridge is of Aymestry Limestone, as is the ridge to the left, in which Leinthall Earls Quarry can be seen. The repetition of the outcrop of the Aymestry Limestone results from displacement on the Leinthall Earls Fault which runs between the two ridges (Fig. 6.15). (© Derek Foxton Archive)

Figure 6.17 (below) Looking west-south-west towards Wigmore and Leinthall Starkes (on the right). During the Ice Age, Leinthall Starkes was on the shore of the old Wigmore glacial lake. The wooded ridge is in Aymestry Limestone which, to the north-east, is a major feature in the Ludlow anticline in Mortimer Forest. The distant wooded ridge to the left is Croft Ambrey, also on Aymestry Limestone. (© Derek Foxton Archive)

of Woolhope (SO 603 387 to 614 381) and a small quarry within Park Wood, north-west of the Wyche cutting (SO 7630 4432) (Fig. 6.18), there are good examples of unbedded masses of reef limestone (patch reefs, referred to by the quarrymen as 'ballstones'). Just over the border into Gloucestershire at Hobbs quarry, east of Longhope (SO 695 194), spectacular examples of fossilised algal reefs (stromatolites) can be seen, surrounded by, and buried by both normally bedded and nodular limestones.

Figure 6.18 Park Wood Quarry, Colwall. This quarry, in the Much Wenlock Limestone Formation, illustrates how patch reefs grew on the floor of the shallow middle-Silurian tropical sea. The bedding would originally, of course, have been horizontal, and it may be visualised that the reef on the right grew on the sea floor, expanding when conditions were favourable, then eventually shrinking in area and dying off – hence the lenticular form in cross section. They do not show bedding because they were composed of a mass of living organisms – corals, algae, bryozoans and stromatoporoids.

Figure 6.19 The Silurian tabulate coral Halysites, photographed in Park Wood Quarry, near Colwall. This was one of the reef-building organisms actively constructing the ballstones during the Silurian. (© Dave Green)

Box 6.2 **The Herefordshire Lagerstätte**

Very rarely, locations are discovered which contain a great diversity of fossil species preserved in very great detail. Such locations are known by the German word *lagerstätten*, literally meaning 'storage places'. A well-known example is the Burgess Shale, of Middle Cambrian age, in the Canadian Rockies. In fossilisation in general, only hard body parts remain but in lagerstätten completely soft-bodied creatures are preserved, showing us something much nearer the full range of the life of the time. The ability to study the internal organs has enabled us to understand so much more about how the animals lived.

At a small site in the west of Herefordshire a lagerstätte was laid down in Silurian times which has made the county world-famous in palaeontological circles.[13] A distant volcano erupted explosively, showering fine wind-borne ashy debris into the moderately deep water at the edge of the Herefordshire shelf (Fig. 6.3). (The location of the volcano is not known but it is possibly connected with the nearest known volcanic deposits of this age, which are in the Dingle peninsula of south-west Ireland.) Creatures living on the sea floor were entombed by the fine ash, which quickly turned to clay (bentonite). As the animals' soft parts decayed, they released carbonate that combined with calcium from the ash to produce calcium carbonate (lime or calcite). This precipitated in the spaces left by the decaying soft tissues, perfectly preserving their detail in resilient calcite. Calcite continued to accumulate around such nuclei to produce rounded nodules, providing further protection for the delicate fossils and preventing squashing by the weight of sediments later laid down on top.

This site, whose location is still secret, was discovered by academic researchers in 1995 and has yielded a rich supply of fossils. Some of the fossils are very good specimens of organisms commonly found in 'ordinary' locations. For instance, the trilobite illustrated in Fig. 6.20 is of good but apparently normal preservation. But there is no other locality in the world, Silurian in age, where trilobite appendages are found. If this specimen were to be treated as described below, it would probably show internal soft body structures as with other fossils from this site. Among the wealth of variety of species only five trilobites have been found at the Herefordshire Lagerstätte.

The many other remarkable finds at this site include a starfish with tube feet preserved, a brachiopod with a pedicle to

Figure 6.20 An example of one of the calcite nodules from the Herefordshire Lagerstätte, showing a trilobite (Tapinocalymene) *preserved inside it. This specimen is held at Hereford Museum. (© John Payne)*

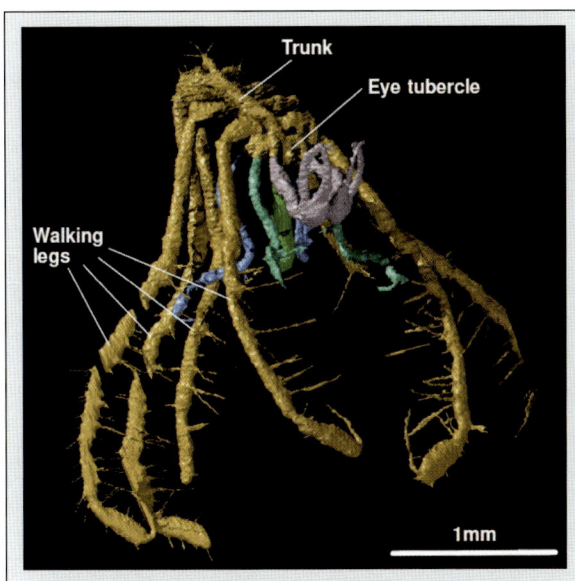

Figure 6.21 Silurian sea spider, Haliestes. Note the preservation of very fine detail, shown well on the legs. (© Dr D.J. Siveter)

tether it to the substrate, a gastropod with soft parts, the sea spider *Haliestes* (Fig. 6.21)[14] and wonderfully preserved radiolarians just a fraction of a millimetre in size. Most recently it has been reported that the oldest parasitic organism has been discovered, a 'tongue worm', attached to a crustacean.[15]

But there are many fossils of creatures that have never been found before, often with amazing detail of their soft parts preserved. It is not possible to remove the fossil from the encasing hard rock but its details can be painstakingly reconstructed by a process which leads to its destruction. A two-micron thick slice is removed from the cut edge of a sawn nodule, the nodule surface is photographed, then another two microns is ground away, and so on – a process known as *microtoming*. The successive images are then put into order in a computer, producing a three-dimensional representation of the complete creature (and even models, using a 3-D printer) with outstanding detail, even down to the earliest known male organ, from a microfossil called an Ostracod. In the *Sun* newspaper on 5th December 2003 this was reported under the headline, 'Old Todger'!

The Herefordshire Lagerstätte is thus a unique feature of our county's geology, perhaps even more valuable than the Burgess Shale in the Rockies as there are very few other examples of Silurian age in the whole world.

Conditions must have been similar over the whole area – shallow, warm and clear sea but without wave activity on the sea floor – the limestones are mainly fine-grained and most of the fossils are unbroken. These reefs were not like the Great Barrier Reef, exposed at low tide and forming a great bastion against the waves but more like the patch reefs growing today in quiet lagoons like Truk, Micronesia, in which many Japanese warships were sunk by US bombers in 1944. The sunken ships were colonised in the years following as 'hard standing' by corals and other organisms, such as microscopic calcareous algae – plant-like organisms living in colonies that build up thin layers of lime, producing mound-like structures called stromatolites. Normally-bedded limestones are frequent and, depending whether they were deposited in agitated or calm water (shown by whether they are coarser or finer grained), contain a profusion of broken or whole fossils (Fig. 6.22).

Another very common rock type of the Much Wenlock Limestone is the strange nodular limestone (Fig. 6.23), consisting of a fine lime-mud matrix from and in which have grown

Figure 6.22 Fossils are often found, mostly as fragments, on the weathered surfaces of bedding planes. This example, from the Much Wenlock Limestone, includes two complete brachiopods, crinoid ossicles, and broken pieces of bryozoans. (The scale has 1cm per division.)
(© John Payne)

Figure 6.23 Middle Silurian Much Wenlock Limestone Formation, here in typical nodular form, photographed in the old quarry on the Knob near Woolhope (SO 604 341). (© Dave Green)

rounded to oval concretions much richer in lime, often enclosing fossils. Some of the best fossils can be found preserved whole in this type of rock, laid down in calm but shallow tropical water.

Both types were well displayed in the quarry at SO 732 363 near Eastnor Castle. During the time it was active, this was a well-known site for trilobites, as well as numerous brachiopods, corals, crinoids and other fossils. Another well-known quarry is Whitman's Hill, near Storridge (SO 748 483), where massive, then silty, then nodular limestones overlie the top of the Coalbrookdale Formation (Wenlock Shale). All are very fossiliferous. There are far too many old quarries to mention them all, but two accessible ones which show the contact with younger beds are Linton Quarry in Gorsley (SO 677 257) and an old quarry (Gurney's) just east of Ledbury (SO 717 383) (Fig. 6.24). Most important of all (although, like Gurney's, not to be used for fossil collecting because of its high scientific status), is the earlier-mentioned Pitch Coppice Quarry in Mortimer Forest (Fig. 6.8).

Towards the west of the county, in the Hergest Ridge/Bradnor Hill area, near Kington, the thickness of the sediments suddenly increases, the sediments change in character, becoming dominantly dark grey siltstones, and the main types of fossil become the swimming, straight (or orthocone) nautiloids (early relatives of today's Pearly Nautilus), and the floating graptolites. Here we are nearing the line of the great Church Stretton Fault, marked today by a series of valleys eroded into the shattered rocks along the fault line to the north-west of Hergest Ridge and Bradnor Hill. The discontinuous, straight line valley

Figure 6.24 Gurney's Quarry, near Ledbury. This peaceful spot, a stone's throw from the town of Ledbury, is an extensive site associated with a nature reserve. The quarry has been cut using the two major vertical joint directions which are at right angles to one another but at about 45 degrees to the direction of the face, producing the buttresses of rock protruding out. It is very easy to see the junction between the underlying Much Wenlock Limestone Formation (here a nodular limestone) and the Lower Ludlow Siltstone Formation above.
(© Dave Green)

Figure 6.25 Bradnor Hill, near Kington, is in the centre of the picture with the Back Brook valley to the right and then the lower slopes of Hergest Ridge. The Back Brook valley leads to Kington and is an old glacial and meltwater channel. (© Derek Foxton Archive)

separates these hills from those formed by the oldest rocks in southern Britain – the 700 million year old gabbros and volcanic rocks of Old Radnor, eroded into Malvern-like hogsback ridges (Fig. 6.5 and Chapter 4). This is a very spectacular place today even though the fault shows hardly any effects due to movement resulting from earthquakes. In Silurian times, however, the fault was very active, its north-west side subsiding, marking the southeast margin of the deep water Welsh Basin. It was the hinge of a tectonically unstable submarine slope reaching down from the warm, sunlit shelf with its limestone reefs into the muddy depths of the basin (Fig. 6.3). Continuing subsidence allowed more sediment to accumulate – hence the thicker strata seen today. The deeper, murkier and colder water did not encourage lime to be precipitated, or animals to flourish on the sea floor, hence the abundance of swimming nautiloids and floating graptolites. From time to time the unstable nature of the slope, combined with earthquakes, caused masses of sediment to start moving down the slope, folding and disrupting the original bedding planes to produce slumped bedding. These rocks are not well exposed in Herefordshire but are presumed to exist, as in neighbouring districts.

The Ludlow Series

The next division of the Silurian, the Ludlow lasting from 427 to 423Ma, saw an initial rise of sea level but the overwhelming trend is towards a retreat of marine influences from Herefordshire. The fall of sea level locally is ascribed to a small elevation of the crust resulting from the plate collision to the north. Figures 6.3 and 6.4 depict the geography at that time. Once again, the main evidence for sea level change comes from the principle of uniformitarianism – as sea level falls in an area, the influence of waves on the sediments on the floor of the sea increases, the water becomes more agitated and the finer material is not able to settle out from suspension. This leads to coarser-grained sediments. These conditions also favour a different kind of more specialised fauna, thriving in shallower, more turbulent water. Eventually, the only organisms to survive are those that can cope with the demanding conditions of the inter-tidal zone, avoiding the dual hazards of water having higher or lower than normal salinity and/or desiccation when the tide is out.

The dominant type of sediment laid down on the shallow water shelf area to the east was silt, laid down much more quickly than lime could be deposited and thus replacing the Wenlock limestones with calcareous (lime rich) siltstone. The coastline must have been some distance away because there is no sign of beach or near-shore sandstones. It is thought, mainly from the evidence of the fossils contained within these beds, that the entire sequence was deposited in water less than 200 metres deep (the typical depth today on the outer part of the continental shelf). These rocks have a very distinctive 'olive' colour due to the presence of the greenish iron-bearing clay mineral, chlorite. They are frequently thin-bedded, indicating rapid changes in environmental conditions. Because of their lime content, making them tough, and their ease of splitting into thin sheet-like blocks, they were frequently used for building stone. They are a characteristic feature of the farms and villages built on their outcrop. The muddier layers are frequently very rich in fossils, especially brachiopods, solitary corals, bivalve molluscs and, sometimes, trilobites and graptolites (Fig. 6.26). Where these have been exposed at the surface for some time, the lime of the fossil shells has been dissolved by modern weathering. This leaves hollow impressions (known as moulds) in the rock, whose iron content has been oxidised brown-yellow by the same weathering to produce a rock sometimes known as 'Gingerbread' or 'Rottenstone'. Individual beds often show little internal structure because they were extensively burrowed ('bioturbated') by organisms such as worms. There is some doubt as to whether the mudstones are actually bedded – some geologists believe that the laterally impersistent and irregular 'bedding planes' are a result of modern upward expansion as the overburden of younger sediments was removed by erosion. The coarser (but still relatively fine-grained) and definitely bedded sandy layers were probably laid down relatively quickly during storms, when wave movement reached down to greater depths. They are frequently lacking in fossils, and are generally

Figure 6.26 A graptolite from the Lower Ludlow siltstones. (The frame width is 6.5mm.)
(© John Payne)

un-burrowed, showing fine laminations and making very good building stone and sometimes flags and roof tiles. At the same time, in some areas the finer sediment was swept away, leaving behind the coarser and denser shells forming a 'coquina'. Over most of Herefordshire these siltstones, mudstones and fine sandstones form two thick sequences, the **Lower** and **Upper Ludlow Siltstone** Formations, separated by the **Aymestry Limestone**.

Topographically, this arrangement, of a relatively strong rock sandwiched between two soft ones, produces classic 'scarp-vale' scenery, in combination with the older Much Wenlock Limestone, Coalbrookdale Formation and Woolhope Limestone. The weaker mudstone/siltstone layers are eroded along the line of their outcrop by rivers and streams (the 'strike' direction), leaving the stronger limestone standing up as ridges or escarpments, again along the strike. Good examples of such ridges include most of the hills to the east of Ledbury, where the Aymestry Limestone comes to the surface several times due to folding, including Sitch Wood and Chances Pitch, to the west of the Ridgeway (Fig. 2.3). A further example is the outer ring of hills around the Ludlow anticline. These are repeated to the south-east in the Leinthall Earls area, this time by a fault (Fig. 6.15), as far south and west as the village of Aymestrey itself. To the north lies the Downton Syncline, where the Aymestry Limestone escarpment runs high above Leintwardine.[16] The limestone re-emerges from beneath the Old Red Sandstone cover to the south forming the outer rim around the Woolhope Dome, well-marked in the north and east where the main TV transmitting mast at Much Marcle rises from its crest. The Aymestry Limestone is extracted at Perton Lane Quarry (SO 595 399; Fig. 6.27). The Woolhope escarpment is much broken by faulting on the west side (Fig. 6.13).

In the north of Herefordshire, the Aymestry Limestone is up to 45m thick (Fig. 6.28)[17] but it thins rapidly to the south of the county, being replaced by a conglomerate of limestone pebbles, suggesting deposition of the limestone, then hardening followed by its erosion. It disappears totally along with all the underlying beds at Gorsley, where the Upper Ludlow Siltstone lies directly on the Much Wenlock Limestone at Linton Quarry (SO 677 257). This suggests that the Gorsley – May Hill area was tectonically uplifted, possibly along the fractures that were to become the Woolhope and Malvern Faults (Fig. 6.3). The instability continued, because the succeeding Upper Ludlow Siltstone is much reduced in thickness and has conglomerate layers, indicating further periods of erosion. The total Ludlow thickness, 350 metres at Ludlow and 400 metres at Woolhope, thins to 60 metres south of Much Marcle and is reduced to a total of four metres at Gorsley, before thickening again at May Hill and further south.

The Ludlow sequence also changes in thickness and rock types as we travel towards the western boundary of the county and the large faults that mark the boundary of the subsiding Welsh Basin. From Woolhope (400 metres), the series thickens to near 600 metres at Leintwardine and to possibly 1,600 metres at Kington, which lay on the basin's eastern slope. Near Leintwardine, the top of the shallow water Aymestry Limestone is suddenly overlain by deep-water graptolite-rich shales and siltstones, indicating sudden deepening of the water. At Mocktree Quarries (SO 417 753) (Fig. 6.29) a section can be seen across a channel, representing the head of a submarine canyon (Fig. 6.3) that was cut by a high density mass of sediment-charged water (a turbidity current) that sped down the

Figure 6.27 (above) This picture is the view south-east over the eastern edge of the Woolhope Dome. The twin wooded ridges of Aymestry Limestone (left) and Much Wenlock Limestone (right) are clearly seen. Perton Quarry, in the foreground, is at the northern edge of the Dome where the limestone outcrops bend to the west (right side of the frame). (© Derek Foxton Archive)

Figure 6.28 (right) The outcrop of the massive Aymestry Limestone in Aymestrey Great Quarry. Above the limestone is the thinly bedded Upper Ludlow Siltstone. This was one of the sites visited by R.I. Murchison with the Revd T.T. Lewis, curate at Aymestrey, during the former's first visit to the area in 1831; and later by E. Alexander during her researches in the 1930s. (© Robert Williams)

slope created by the fault movement. Modern examples have been measured as travelling at speeds of up to 100km/hr, and are thought to be capable of rapid, deep incision of gullies into the continental slope. The fault movement would also have generated earthquakes that disturbed and stirred the sediment into suspension. The material filling the channel is much younger than the sediments in the walls and floor of the channel and it buried a well-preserved fauna that colonised the channel floor. This includes starfish, eurypterids (extinct crustacea), fish and crinoids, as well as more typical brachiopods, trilobites and graptolites, which were found particularly at the nearby Church Hill Quarry (SO 412 738).

On the slope itself, to the west, partly consolidated sedimentary layers, mainly of often bioturbated mudstone and siltstone, were disturbed by earthquakes and the cutting of the canyon slopes. They slid down the slopes, often rumpling and folding rather like a cloth slipping off a table. Excellent examples of such features can be seen at the junction of the spectacular Sned Wood and Covenhope fluvio-glacial channels, in a small quarry (SO 402 653) on the western edge of Mere Hill Wood (Fig. 6.30). The area of Herefordshire to the south, around Kington and comprising Bradnor Hill and Hergest Ridge, has not yet been geologically surveyed in detail but is thought to be of the same nature.

Figure 6.29 Mocktree Quarries, near Leintwardine, show some of the features associated with the erosion of submarine canyons as the sea floor to the west deepened, creating an underwater slope down which sediment slid and was transported by turbidity currents. The photograph shows a section across the shallow head of such a canyon; nearby the channel deepens and cuts a vertical slope against Lower Ludlow Siltstones. The canyons appear to have been up to 125 metres deep.
(© John Payne)

Towards the top of the Upper Ludlow Siltstones, there is much evidence that the water was markedly shallowing all over the county area. The thin-bedded shales and siltstones of the outer shelf area to the west give way to thicker-bedded, coarser-grained flaggy, silty and sandy sediments. In the east, in amongst the bioturbated siltstones, there is an increasing amount of thin-bedded sandstone deposited during storms, when the wave base was lowered. In the south, around the already very shallow area of Gorsley, there are several phosphatised pebble beds, rich in fish bones, scales and teeth and deposited as 'lags'. The pebbles are the only material with enough mass to resist being swept away by wave and current action in very shallow, possibly intertidal, water; those people who have looked along the high tide strand line for dense pyritised fossils washed out of mudstone cliffs at beaches such as Charmouth in Dorset will recognise this idea. Another common lag deposit is the streaks of black minerals found on sandy beaches, which consist of heavy minerals, such as magnetite and ilmenite. The phosphate is probably derived from the excrement and bony material from vertebrates, particularly fish (bone is calcium phosphate). As far as invertebrate fossils are concerned, the fauna is dominated by brachiopods, particularly *Salopina* and *Protochonetes*, which occur often in massive numbers, but with little else in the way of diversity.

Figure 6.30 Mere Hill Wood Quarry, at the north-west end of the Covenhope Valley. This old quarry lies in the dramatically atmospheric landscape of a valley deeply incised by glacial meltwater along the line of the Leinthall Earls Fault. It shows Lower Ludlow Siltstones, here much thicker than further east due to contemporary subsidence on the nearby Church Stretton Fault system. The quarry shows a great thickness of normally bedded siltstones and mudstones, with at least one prominent slumped bed that must have slid down the submarine slope as a semi-consolidated sheet, rather like a tablecloth sliding off a polished table, rucking and folding up on itself. Some of the bedding planes have been highlighted to illustrate this. (© Dave Green)

The Přídolí Series

All across the county the base of the last of the divisions of the marine phase of the Silurian is defined by the Ludlow Bone Bed Member, the most widespread of the lag deposits mentioned above. It contains abundant vertebrate remains, phosphates and the carbonised remains of early plants such as *Cooksonia*. Good exposures of the Ludlow Bone Bed are small, delicate and have been much disturbed by fossil collectors so that they are rarely seen today.

Above lies the *Platyschisma* (now *Turbocheilus*) Shale Member, with abundant specimens of this gastropod (sea snail), followed by local representatives of the Downton Castle Sandstone Formation, usually yellowish in colour, laminated and often cross-bedded. It contains a sparse, restricted fauna of mussels, gastropods and the intertidal brachiopod *Lingula*, besides the fish, eurypterid (Fig. 7.12) and plant fragments. It represents the sandy shoreline that advanced rapidly westward as the water shallowed and was replaced by the red terrestrial deposits of the Old Red Sandstone deposited on a wide coastal plain. In places it is durable enough to constitute an important building stone and has been extensively quarried, for example, in the type area of Downton Gorge, west of Ludlow but still just in Herefordshire such as at location SO 453 750 (Fig. 6.31).

There were extensive quarries on Bradnor Hill near Kington (such as SO 291 577, near the golf clubhouse). In the south, it was quarried

Figure 6.31 An old quarry in Downton Castle Sandstone near the bridge over the River Teme at Bringewood Forge (SO 453 750). The lower part of the quarry is in the sandstone, used extensively in the area as a building stone, including the nearby bridge. At the top of the quarry lies the green-grey/red basal part of the mudstone of the Moor Cliffs Formation (Raglan Mudstone) – the base of the Old Red Sandstone – indicating the change from a marine to a terrestrial environment. The wide joint spacing and relatively thick beds of the sandstone produce large unflawed blocks. This, combined with its resistance to weathering, made it a much sought after building or dimension stone (for working to specific sizes). (© Dave Green)

as 'Gorsley Stone', in large workings in and around the village, and can be seen well exposed at the top of Linton Quarry (SO 677 257), as well as in numerous buildings, where it weathers to a pleasing pinkish yellow colour.

In the south-east of the county, brown to yellow-brown sandstones are interbedded with argillaceous strata on a less than one metre scale and are best exposed adjacent to and in the Perton Lane Quarry (SO 595 399). Greatly fragmented eurypterid fossils are commonly present and well preserved specimens of *Eurypterus brodiei* were found by several mid-19th century geologists including Brodie[18] and Woodward. Articulate Ludlovian brachiopods survive into the lowest 25cm of the Rushall Formation around the Woolhope Dome and so the rock is taken to equate with the main Downton Castle Sandstone in the north (Fig. 6.10).

The red mudstones of the Moor Cliffs Formation (Raglan Mudstone) lie above the Downton Castle Sandstone. These rocks are formally included in the Silurian (because this fits better in a worldwide context, defined by the fossil content) but are not of marine origin so are most sensibly included, in this local description, in the Old Red Sandstone, which consists of the subsequent terrestrial rocks. The Old Red Sandstone includes also all of the Devonian rocks and is described in the following chapter.

The Přídolí division of the Silurian is named from a locality in the Czech Republic and the Moor Cliffs Formation from Pembrokeshire. Otherwise, all of the formal names used in this book for the divisions of the Silurian are of local, Welsh Borderland, origin. This, with the presence in Herefordshire of two sites of worldwide importance, the lagerstätte and the Pitch Coppice stratotype, emphasises the great significance of this area in Silurian geology.

Further reading

The widely separated nature of the Silurian outcrops around the county necessitates a reading list based to some extent on the different areas.

A book which covers the whole area is *The Welsh Borderland* in the British Regional Geology series, published by British Geological Survey. Despite its age (1971), it contains a great deal of useful information.

There are relevant sections/chapters in the following books:

May Hill / Malverns: in William Dreghorn's *Geology Explained in the Severn Vale and Cotswolds*, (1967), reprinted several times, the latest being 2005, by Fineleaf Press.

Kington to Old Radnor: in *Geological Excursions in Powys*, ed. N.H. Woodcock and M.G. Bassett, University of Wales Press (1993).

Mortimer Forest (Wigmore Dome): 1. Geologists Association Guide No.27 *The Geology of South Shropshire*, ed. J.T. Greensmith (2002).

Mortimer Forest Geology Trail, ed. Andrew Jenkinson, Scenesetters/Forestry Commission (1991).

The multiple relevant publications of the Earth Heritage Trust and the British Geological Survey are listed on the page 'EHT and BGS Publications'.

One of the major points of interest in the marine Silurian rocks described in this chapter is the abundance of fossils. Although there are illustrations in many of the above books, the best source of information to help identify the majority of the fossils you may find is the relevant volume in the long running, beautifully illustrated series published by the Natural History Museum: *British Palaeozoic Fossils*, fully updated in 2012 as *British Fossils: Palaeozoic*.

Also see the Geological Conservation Review, vol. 19 (2000) *British Silurian Stratigraphy*, Aldridge, R.J. *et al.*, vol. 9 (1995) *Palaeozoic Palaeobotany of Great Britain*, Cleal, C.J. & Thomas, B.A., vol. 16 (1999) *Fossil Fishes of Great Britain*, Dineley, D. & Metcalf, S. and vol. 35 (2010) *Fossil Arthropods of Great Britain*, Jarzembowski, E.A. *et al.*

7 THE OLD RED SANDSTONE – ALL OF HEREFORDSHIRE ABOVE SEA LEVEL

> ... there is no region of Europe yet examined where the Old Red Sandstone is better exhibited than in the British Isles ... [and] ... there is no part of the kingdom in which it is so much expanded as in the country here described. Occupying the largest portion of Herefordshire and the adjacent districts of Worcestershire and Shropshire, it spreads over wide tracts of Monmouthshire, surrounding the coal-field of the Forest of Dean and forming a girdle round the great South Welsh coal basin, it constitutes in Brecknockshire the loftiest mountains of South Britain. ... in the central districts of Herefordshire the strata lie in a great basin, the lower edges of which are turned up against Silurian rocks, both on their eastern and western flanks.
>
> Sir Roderick Murchison, *The Silurian System* (1839) p.170

Introduction

With the exception perhaps of the Black Mountains foothills above the Olchon valley, Herefordshire's green and pleasant land successfully hides its geological foundations. From its highest upland plateau to its central water meadows, Herefordshire is very largely a county of Old Red Sandstone (ORS) (Fig. 1.3), which controls its characteristic scenery of undulating ridges and its rich clay soils that break down to a fertile red loam.

In fact, Herefordshire (Fig. 1.2) is more a landscape of water meadows than steep mountainsides and it is on some of the major watercourses that tantalising glimpses of the underlying rocks can be seen. Riverside sections at Breinton (SO 4518 3994) and at Redbank Cliff, Holme Lacy (Fig. 7.2; SO 5560 3614) are key sites for some of our earliest Old Red Sandstone sediments, but the most spectacular is Brobury Scar. Cutting past Brobury and Monnington-on-Wye, the fast-flowing River Wye exposes, on the outside of a large meander loop, 30m of the topmost Moor Cliffs Formation (previously known as the Raglan Mudstone Formation; Fig. 7.3). This 'Brobury Scar' (Frontispiece page *xiv*; SO 354 444) is largely inaccessible and is best seen from across the river at Deepwell. It was the subject of a lovely drawing by Thomas Hearne, the Picturesque artist[1] and was once described as 'one of the biggest as well as one of the most beautiful exposures of the Old Red Sandstone in Herefordshire'.[2] Although somewhat overstated, this comment underlines the importance of these Wye river sections in unravelling the geology of our county.

The 'Devonian System' was first named by Murchison and Sedgwick in 1839 to describe the fossiliferous marine rocks of both Devon and Cornwall. This geological period lasted

Figure 7.1 Herefordshire's south-western boundary runs north-west along the crest of the Black Mountains ridge. To the west (left) in Wales is the valley of the Afon Honddu. To the east is the Olchon Valley, running up to Black Hill (the Cat's Back, Fig. 2.8) on its eastern side. The two nearby promontories in this major Old Red Sandstone escarpment are Black Darren (the nearer) and Red Daren (Chapter 9). (© Derek Foxton Archive)

from 419 to 359Ma and it includes much but not all of Herefordshire's Old Red Sandstone. 'Old Red Sandstone' has always been a useful term in both south Wales and the Welsh Borders as the onset of a red colouration in the rock types coincided with the change from fully marine to river-based (fluviatile) conditions. With fossiliferous marine Silurian conditions continuing longer in many areas to the south, together with their abundant graptolite faunas, some older parts of the former Old Red Sandstone are now seen as contemporary with these marine strata and so must be included in the Silurian system. Hence the Silurian-Devonian boundary, discussed below, falls within our ORS sequence and not at the base of it; the Devonian Period does not include the lowest part of the Old Red Sandstone.

Origins of the Old Red Sandstone

Starting in the late Silurian (Přídolí stage), and even earlier in Scotland, there was a gradual shallowing of the continental shelf seas of the Iapetus Ocean which had dominated geological events in the Welsh Borders since Cambrian times (Fig. 3.9). This shallowing was a precursor to the eventual closure of this ocean by collision with the northern continent of Laurentia and the resulting formation of a series of mountains of probably Alpine proportions as the first phase of the Caledonian orogeny.

This closure event essentially brought together the three main components of the Old Red Sandstone continent (Fig. 3.11). The old continent of Baltica had collided with

Figure 7.2 A view of the mudstone of the Moor Cliffs Formation at Redbank Cliff (SO 555 361) taken from the opposite bank of the River Wye at Holme Lacy. (© John Stocks)

Eastern Avalonia closing the Tornquist Sea and had then moved northwards to collide with Laurentia in late Silurian times, completely closing the Iapetus Ocean. The Old Red Sandstone has long been seen as the first product of the erosion of the emergent Caledonian mountains. These lay to the north of Herefordshire, in north Wales and northward. Our county then was near the southern margin of the new continent. Large quantities of sediment arrived on the continental margin from the north through fast-flowing rivers which developed complex deltas with distributaries and back swamps as they entered the Rheic Ocean to the south.

The Devonian Period continued to be one of active plate tectonics after the formation of the Old Red Sandstone continent, with the development of active mid-ocean ridges in the Rheic Ocean. This process, through the consequent reduction in ocean volume, led to the displacement of seawater into continental areas and major marine transgressions are noted particularly in the late Devonian. Part of this story may be explained by the fact that the number of palaeomagnetic reversals during the Devonian was particularly low, indicating essentially a 'quiet' magnetic period in terms of the Earth's core.[3] This suggests that the core may have lost heat to a series of dynamic mantle plumes which were responsible for generating the earlier high levels of plate tectonic activity. Palaeomagnetic reconstruction (Box 3.1) places the Old Red Sandstone continent in tropical to sub-tropical latitudes with the Welsh Borders sitting somewhere between 5° and 15° south of the equator.[4]

Palaeomagnetic studies on the Lower Old Red Sandstone further west, in south Wales, produce a palaeolatitude of 17 ± 5° south which is in broad agreement and supports the idea of the hinterland being arid and the consequent production of reddened sediments rich in the mineral haematite (iron oxide). Wet and dry seasonal variations are also seen at similar latitudes today and could give rise to the observed calcrete palaeosols, described below. Such palaeosols (fossil soils) are typically produced in a warm to hot, arid or semi-arid, tropical to subtropical climate. Rainfall in such areas is confined to a wet season and thus can be equated with a monsoonal-type environment.[5]

These non-marine deposits, carried from the northern mountains and laid down on an alluvial plain, contain fossils of primitive freshwater fish and a sparse land flora. Along the Welsh Borders, the identification of stratigraphic boundaries within the Old Red Sandstone has always been hampered by the poor preservation of both its floral and faunal elements. Correlation with the abundant ammonoid, conodont and late graptolite assemblages of south-west England, Belgium, Germany and Bohemia has proved to be very difficult and

Divisions in this book	Formal divisions	Series names	Approximate Thickness / m	Current Formation Names	Old Formation Names (pre-2015)
Old Red Sandstone (Terrestrial)	Devonian	Upper ORS	75	Tintern Sandstone Fm	Tintern Sandstone Fm
			10	Huntsham Hill Conglomerate Fm	Quartz Conglomerate Fm
		Middle ORS		Missing in Herefordshire	Missing in Herefordshire
		Lower ORS	1200	Brownstones Fm	Brownstones Fm
			300	Senni Fm (only on Black Mountains)	Senni Beds (only on Black Mountains)
			700	Freshwater West Fm	St Maughans Fm
			15	Chapel Point Limestone Fm	Bishop's Frome Limestone Fm
			30 / 20	Moor Cliffs Fm	Raglan Mudstone Fm
	Silurian		3	Townsend Tuff Bed	Townsend Tuff Bed
			600	Moor Cliffs Fm	Raglan Mudstone Fm
			20	Temeside Mudstone Fm (only in north Herefordshire)	Temeside Shale Fm (only in north Herefordshire)
Marine Silurian		Přídolí	15	Downton Castle Sandstone Fm	Downton Castle Sandstone Fm

Figure 7.3 A representative succession of the Old Red Sandstone within Herefordshire to show the relationship between the Old Red Sandstone, the Devonian and the Silurian strata. Biostratigraphic analysis places the Silurian-Devonian boundary 30 metres below the Chapel Point Limestone Member.[6] The names of many Formations of the Old Red Sandstone were changed by the British Geological Survey.[7] The new names are used in this book and their correlation with the old names is shown in the table. Only the main Formations are shown here. The Formation thicknesses in the table are indicative only; they vary widely across the county.

remains one of the key challenges for today's geologists working on the Old Red Sandstone. Spore analysis offers a way forward to correlate the marine rocks elsewhere with our continental facies and recent work now places the important Silurian/Devonian boundary within the Moor Cliffs Formation (Raglan Mudstone), 30 metres below the Chapel Point Limestone Member (Bishop's Frome Limestone; Fig. 7.3). This defines both the Downton Castle Sandstone Formation and most of the succeeding Moor Cliffs Formation as being of late Silurian (Přídolí Stage) and equivalent to the highest graptolite zone (*Monograptus uniformis*) found in deep-water mudstones of the Czech Republic.[8]

The names of the formations making up the Old Red Sandstone were reviewed by the British Geological Survey[9] and several were changed, with some names from the Welsh Borderlands being replaced by ones from Pembrokeshire, where there are excellent exposures in the sea cliffs. The correlations are shown in Figure 7.3. In addition, this table shows how the Přídolí Stage of the Silurian System forms also a part of the Old Red Sandstone. As mentioned earlier, the Old Red Sandstone is an informal unit convenient for use in this area and in general use throughout the United Kingdom.

The Downton Castle Sandstone Formation (Marine Silurian)
The start of the Přídolí Stage late in the Silurian Period marks the end of the marine conditions and the final infilling of the Anglo-Welsh Basin on the edge of the closing Iapetus Ocean. The expected sandy shoreline deposit is represented by the Downton Castle Sandstone Formation, discussed in Chapter 6, in the north of the county and by the 'Gorsley Stone' of the Ledbury area. Around Hereford and the south-east of the county these strata are referred to as the Rushall Formation. This rock is essentially of marine origin and so is regarded as the topmost formation of the marine Silurian rocks. Its top represents the transition to river-based (fluviatile) deposits and so is now taken as the base of the Old Red Sandstone.

The Moor Cliffs Formation
The lowest beds of the Old Red Sandstone are chiefly mudstones. (Note that these beds are still formally included in the Silurian System because of their fossil spore content.) Alluvial plain deposits characterise these strata and are designated as the Temeside Mudstone Formation in the north and west of the county. Here a series of olive-green massive mudstones with thin horizons of calcrete nodules are indicative of periodic desiccation events.

The Temeside Mudstones are followed by the Raglan Mudstone Formation. Both are included in the newly named Moor Cliffs Formation. In the rest of the county the Moor Cliffs rocks directly succeed the Downton Castle Sandstone and underlie much of the central Herefordshire plain, so being the origin of much of the county's red soil. These easily erodable red mudstone sediments are rarely exposed except as river cliffs within meander loops as described earlier. To the west of the River Lugg they are buried beneath Devensian glacial deposits. Only 10% of the Formation consists of sandstones, whose additional content of grains of both feldspar and mica hints at a granite source within the former catchment of the supplying streams. In addition, there are smaller quantities of rarer and

Figure 7.4 Moor Cliffs rocks at Westonhill Wood (SO 314 453). (© John Payne)

heavier mineral species, primarily garnet but also epidote, apatite and tourmaline which could only have originated from metamorphic and igneous rocks within the Caledonian mountains. Scottish and northern Irish sources are strongly suggested by the direction of river flow at the time, shown by the orientation of cross-bedding within these thin sandstones. The Acadian Orogeny of the mid-Devonian uplifted many of the deposits from the early Devonian, leading to their erosion. The loss of these rocks has severely restricted our picture of the whole sedimentary basin.

The general nature of the landscape in our area at this time is shown in Fig. 7.5. A collision from the south – the Acadian Orogeny – drove the Midland Platfom (now central England) north-westward against the Welsh Basin. The sedimentary pile in the Welsh basin was thickened by folding and began to rise, forming a new land area. Rivers from the uplifted region (the forerunner of the landmass sometimes known as St George's Land) flowed south across what is now Herefordshire but was then an arid to semi-arid desert bordering a shallow sea to the south. As the land rose, swampy coastal plains were replaced by meandering rivers, followed by braided sandy river systems, and finally became transformed into a low lying landscape that had stopped subsiding and therefore preserved no sediment.

Primitive plants occupied the ground close to water. The trace fossils of plant roots are not uncommon through most of the ORS rocks although they are often not easily recognised. Figure 7.6 shows the cavity left by a decayed root, later filled by mud.

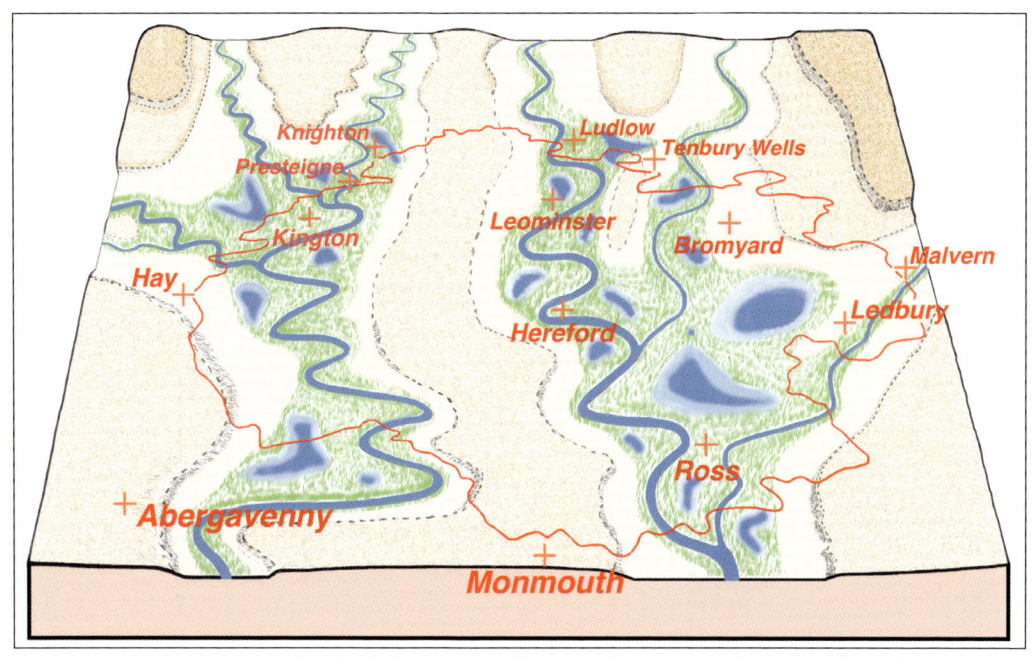

Figure 7.5 A figurative sketch of the Herefordshire landscape in early Old Red Sandstone times (c.400Ma). The position and form of the rivers shown is purely conjectural. (Drawn by Gerry Calderbank from a conceptual sketch by Dave Green)

Figure 7.6 A trace fossil (infilled cast) of a plant root in the Moor Cliffs Formation. (© John Payne)

Calcretes

The sandstones occupied the distributaries (secondary side channels of the rivers) within the alluvial plain while the intervening back swamps, where flood deposits settle, were the location for the development of cornstones (immature calcretes) and occasional thicker bands of limestone (mature calcretes). The origin of calcretes lies in the soil-forming process at or near the surface of the water table. Evaporation takes place at the sun-baked surface of the back swamp encouraging the development of networks of desiccation cracks (or sun cracks) and initiating capillary action which brings to the surface deeper groundwater containing dissolved lime. This is precipitated out as calcareous nodules along sub-vertical branching tubes locally called 'race' (which was formerly used as a cheap fertilizer in lime-poor areas of Herefordshire). Closely similar deposits form today in tropical areas with seasonal rainfall. Long periods of exposure allow the development of a mature calcrete layer such as the Chapel Point Limestone Member,[10] named after its type locality (where its best and most typical developed form is seen) on Caldey Island, off Tenby, in Pembrokeshire.

The Chapel Point Limestone was formerly called the 'Psammosteus' Limestone from its associated fish remains before its non-biological origin was recognised. Later it was known as the Bishop's Frome Limestone. It consists of several stacked calcrete profiles totalling about 8 metres in thickness. Somewhat perversely, the fish remains, now designated as two species of *Traquairaspis*, are not in the calcretes but in the intervening sandstones and shales. The formation of the mature calcrete sequence probably took well over 10,000 years and represents a major break in deposition within the Anglo-Welsh Basin.[11] Closely similar calcretes form today in lime-rich soils within both arid and semi-arid areas such as Israel and the south-western USA. Typical annual temperatures and rainfall need to average 16 to 20°C and 100 to 500mm. The Chapel Point Limestone in an immature form may be seen in the bank of a track on Westonhill Wood (Fig. 7.7).

Lower down the hill, the trackside shows good exposures of the Moor Cliffs Formation. (Fig. 7.4) The outcrop of the Chapel Point Limestone has been mined for building stone and for lime at Credenhill (SO 4459 4485) and at Garnons Hill near Bishopstone to the west of Hereford.

Figure 7.7 Chapel Point Limestone at Westonhill Wood (SO 310 454). At the top is an erosion surface with sandstone of the Freshwater West Formation above.
(© John Payne)

Two other rock types seen within the Moor Cliffs Formation were formed on a much shorter timescale; intraformational conglomerates and the Townsend Tuff.

The first of these arose from torrential flash floods, emanating from the Caledonian mountains to the north, which produced cornstone conglomerates. In these conglomerates, the rocks of the desiccated alluvial plain including calcrete layers were rapidly eroded to form a mélange of sub-rounded pebbles set in a calcareous sandy matrix. Good fossil fish specimens are often found in these horizons. The same periodic floods also led to overbank inundation of the intervening back swamps. On wetting, sun-cracked surfaces yielded flakes of mudstone which were then carried along and finally deposited downstream as intraformational conglomerates. These too have a sandy, current-bedded matrix.

The Townsend Tuff

The Chapel Point Limestone defines the top of the Moor Cliffs Formation. Locally, just less than 100 metres below this level, outcrops the Townsend Tuff Bed, the thickest and most widespread airfall tuff (solidified volcanic ash) in the Anglo-Welsh Basin. Present in Cusop Dingle (SO 2486 4026) close to Hay-on-Wye and behind a waterfall on the lower slopes of Merbach Hill (Fig. 7.8), the former pyroclastic deposit is now represented by two bands of porcellanite and an intervening tuff formed from minute crystals.[12] Porcellanite is a compact, dense, siliceous rock with the appearance of unglazed porcelain, often formed by the lithification of very fine volcanic ash.

The Townsend Tuff was first identified on the foreshore at Townsend close to Dale village in Pembrokeshire and has been located over large areas of the Anglo-Welsh Basin. These tuffs, produced by a succession of major Plinian-style eruptions, and named after the AD 79 eruption of Monte Somma (pre-Vesuvius) documented by Pliny the Younger, act as a valuable marker horizon in the absence of strong palaeontological control. Elsewhere in Herefordshire it is either absent or represented by bentonitic mudstones such as those at Hunderton (SO 4862 3876). Further exposures are seen at Bosbury and in the Breinton Gorge west of Hereford but the full picture of the airfall distribution is still incomplete. The location of the source volcano is unknown but is believed to lie somewhere to the west of the outcrop area, which extends to Pembrokeshire.

Figure 7.8 The Townsend Tuff in a waterfall on Merbach Hill (SO 3068 4516). (© John Payne)

Figure 7.9 Sandstone of the Freshwater West Formation in a quarry in Westonhill Wood (SO 310 452). (© John Payne)

The Freshwater West Formation (St Maughans Formation)

The succeeding Freshwater West Formation (Figs. 7.3 and 7.9) is unquestionably of Devonian age and, like the Moor Cliffs Formation, displays cycles of sedimentation. However, sandstones now account for a much greater proportion (35%) of the thickness. The occurrence of metamorphic mineral species has declined sharply within the heavy fraction of this Formation. The 'heavy fraction' consists of the minerals of high density (often iron-bearing) as opposed to minerals such as quartz. The source rocks were again towards the north-west as indicated by palaeocurrent measurements on the cross-bedded sandstones and by this time were mainly sedimentary and volcanic in origin, probably from within the Welsh Massif itself. The presence of increased apatite, tourmaline and zircon and the decrease in garnet grains indicates the changed nature of the source rocks.

The Freshwater West rocks are more resistant to erosion than those of the Moor Cliffs Formation and protect the softer, older ORS formations in many of Herefordshire's upland areas such as the hills of the Golden Valley, the Bromyard Downs and the centrally located range of tabular cornstone hills stretching from Dinmore through to Wormsley. The latter range, including outlying hills at Credenhill, Garnons Hill and Merry Hill, represents an area dissected by ice movement during the last, Devensian, glaciation. Limestones of calcrete derivation occur in the Freshwater West Formation at about 350 metres above its

Figure 7.10 Porch at the church of St John the Baptist, Mathon (SO 734 458) restored in 1897 with detail of the intraformational conglomerate on the right used for the low walls beneath the timber frame. (© John Stocks)

base. This so-called Hackley Limestone outcrops in a discontinuous arc to the south-west of Bromyard (near Hackley Farm, SO 636 534).

The fluvial deposition within the Freshwater West Formation frequently shows a succession of fining-upwards cycles. Upward-fining is a feature often displayed by detrital sediments whereby the maximum grain size progressively decreases from the base to the top of a particular bed. The sequence of sediments showing these features is referred to as a fining upwards cycle. Such deposition occurs as a river channel gradually fills with deposited material, for as it does so, water flow becomes slower meaning that only increasingly fine sediment can be carried in it and then deposited.

Each fining-upwards cycle within the Freshwater West Formation typically begins with an intraformational conglomerate (Fig. 7.10) deposited on the eroded mudstones of the previous cycle. These can be up to one metre in thickness and are rich in calcrete clasts, siltstone flakes and fish remains. Close to Arthur's Stone, above Dorstone, acanthodian and heterostracan fish fragments were extracted from a disused quarry (SO 3145 4297).[13] Large tubular burrows, circular to elliptical in cross-section, have been found in bedding planes within the disused Linton Tile Works near Bromyard (SO 6671 5388) and in the sandstones on Westonhill Wood. These trace fossils (Fig. 7.11), known as *Beaconites antarcticus*, seem to have preferred to live in and close to active river channels but the exact nature of the animal has not yet been determined.[14]

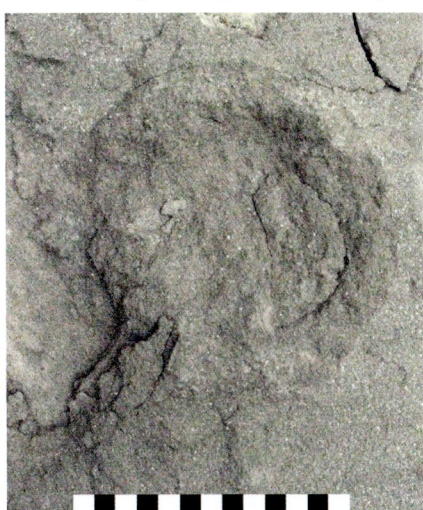

Figure 7.11 A Beaconites trace fossil (the top end of its burrow) in the Freshwater West sandstone of Westonhill Wood quarry. (Scale division = 1cm) (© John Payne)

The Freshwater West Formation, in common with the Moor Cliffs Formation shows fining-upwards cycles, starting with coarse intraformational conglomerates (similar to that of Fig. 7.17) at the base, through sequences of fluvial sandstones, to overbank mudstones and associated calcretes at the top. The higher percentage of sandstone indicates that the sediment sources were closer than previously.[15] The sandstone units often form small waterfalls on local stream courses, especially in the Bromyard Downs. Subsequent erosion, caused by splash pool effects at the base, erodes away the softer mudstone. This reveals the underside of the sandstone units and, sometimes, the preservation of the perfectly formed casts of desiccation cracks, such as at Little Cowarne (SO 6001 5128). Such underlying features are best seen with the judicial use of a mirror fixed to a long stick!

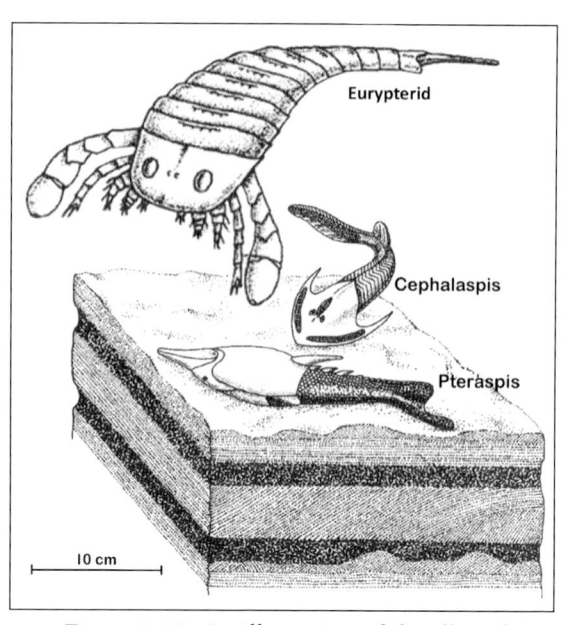

Figure 7.12 An illustration of the alluvial environments typical of the Lower Devonian in Herefordshire. The most common fish are species of Agnatha (such as Cephalaspis and Pteraspis) together with eurypterid arthropods (adapted from Dineley 1984). (Reproduced courtesy of Palgrave Macmillan)

Fish fossils (see also Box 7.1)

The earliest fish evolved in Old Red Sandstone times. Fragments of heterostracan (armoured bodies) pteraspid fish, osteostracans (armoured heads) (including cephalaspids) and scattered thelodont scales are found throughout the Přídolí and later sequences of the Anglo-Welsh Basin (Fig. 7.12). These bear witness to the very high diversity of jawless fishes (Agnatha) found within the Lower Devonian (Box 7.1). These vertebrate remains are joined by a shallow marine to brackish water fauna of ostracods, bivalves, eurypterids and inarticulate brachiopods (lingulids) in the lowest Downton Castle Sandstone Formation. Gradually these invertebrate groups became scarcer towards the top of the Moor Cliffs Formation due to the

Figure 7.13 An acanthodian fish spine (white, on right) preserved within greenish sandstone of the Freshwater West Formation exposed in a river cliff at Lydney (SO 653 017) on the Severn Estuary. (© John Stocks)

increased overall aridity within the Anglo-Welsh Basin. It is still uncertain whether the agnathan (Fig. 7.12) and acanthodian fishes (Fig. 7.13) were solely freshwater dwellers or could equally survive within brackish water environments caused by salt-water incursions extending up low gradient river courses. The fish, like some modern species, could have lived in marine to brackish conditions and then migrated upstream to spawn.

The highest levels of marine influence are seen in the lower 60 metres of the Freshwater West Formation, just above the Chapel Point Limestone.[16] In a former railway cutting at Ammons Hill (SO 700 529), the presence of ostracods, gastropods, bivalves, eurypterids and pteraspid fish fragments suggests that, despite the onset of fully continental conditions, the delta complex was still breached by influxes of salt water. Fully marine conditions existed further to the east at this time[17] and a marine embayment probably extended northwestwards towards the Worcestershire–Herefordshire border to account for this fauna.

| Box 7.1 | **The Age of Fishes** |

After the discovery of an abundance of fossils in the Silurian rocks of the county, early Woolhope Club members, searching in the 1850s, found the Old Red Sandstone frustratingly devoid of fossils. The Devonian Period, a major part of the Old Red Sandstone time, has long been designated as the 'Age of Fishes' as it is then that fish become widespread and the most advanced vertebrate life forms on the planet. Most fish fossils are found as teeth, scales or spines, with whole specimens being rather rare, and it was these scattered finds that first drew the attention of the early fossil collectors. The work of Miller in the 1840s and 1850s on Scottish Old Red Sandstone fossil fish only increased the disappointment of Woolhope members. His spectacular drawings of whole fossil fish found within the much more fossiliferous sequences in Caithness and Orkney were particularly galling. More recently, **ichthyoliths** (fish microfossils) have greatly added to our knowledge of the evolution of fish but the groundwork was already in place from the 19th century research.

Jawless **agnathans**, resembling modern lampreys and hagfishes, can now be traced back to the Ordovician Period around 470Ma. During the succeeding Silurian and into the Devonian, one group of jawless fish (**ostracoderms**) developed external bony head shields and poorly ossified internal skeletons. These groups are represented by the cephalaspids and pteraspids found in the Lower Old Red Sandstone successions and have dorsally-positioned eyes and a mouth on the underside of the head. This suggests that these early fish were bottom feeding animals (Fig. 7.12). The first jawed fish, the **gnathostomes**, are represented in the Anglo-Welsh Basin by the **acanthodians** or spiny-skinned fish. These had a cartilaginous skeleton, similar to sharks, but differed in having a skin in which small bony plates were arranged.

Also present, alongside ostracoderms and acanthodians, were other primitive agnathans, the **thelodonts**. These fish had a well-developed tail with both dorsal and anal fins. Rudimentary pectoral flaps suggest that they were strong swimmers in open water rather than bottom dwellers. The mouth, instead of being on the ventral side, faced forward but the lack of teeth still suggests that they filtered plankton or sucked in small organisms as they moved through the rivers of the Old Red Sandstone alluvial plains.

The absence of bony head shields indicates that speed rather than armour was their main means of defence. The lack of armour in both the acanthodian and thelodont groups also means that fossils of these types found in Herefordshire are limited to scattered fin spines (Fig. 7.13) and bony skin denticles.

Thelodonts, along with spore analyses, are important in defining the Silurian / Devonian boundary with the incoming of *Turinia pagei* at the base of the Ditton Group, a level about 30m below the Chapel Point Limestone on Brown Clee Hill, Shropshire.

The Chapel Point Limestone sequence is a key horizon in the story of the evolution of fish. Below this horizon, freshwater heterostracan ostracoderms such as *Traquairaspis symondsi* and acanthodians dominate the fauna. However, above the main limestone, which is itself unfossiliferous, the jawless fish fauna is dominated by a pteraspid, *Protopteraspis gosseleti*, whose gross morphology suggests a more nektonic (swimming) lifestyle. Only armour plates from adult fish have been found at this level, which may indicate that the juvenile development was in marine waters elsewhere with migration into the rivers of the Old Red Sandstone in later life.[18]

There was, therefore, a major environmental change at this level to which purely freshwater forms were unable to adapt. Only the thelodonts, adapted to both brackish and freshwater lifestyles, are found throughout the sequence from Late Silurian through to the Devonian.

The Senni Formation

Overlying the Freshwater West Formation, but only at the far south-west edge of the county, on the Black Mountains, is the Senni Formation (Figs. 2.8 and 7.14). A thin mature calcrete, the Lower Ffynnon Limestone, at the base of the Senni Formation is followed by 213m of greenish to brown flaggy sandstones which show strong channel development. The uppermost beds again feature a mature calcrete, the Upper Ffynnon Limestone, and several conglomeratic cornstone layers. These beds form a distinctive outcrop on the Cat's Back (or Black Hill; SO 280 338) where the harder sandstone bands form most of the steps on the ridge as it rises towards Black Hill. One very prominent step exposes one of the main calcrete horizons.

Figure 7.14 An example of soft-sediment deformation in the sandstone of the Senni Formation at SO 2830 3350 on the Cat's Back. This deformation occurred, before consolidation of the deposit, in the unstable situation when a layer of wet sand overlay another layer which contained a greater proportion of water, was therefore less dense and so rose through the denser layer.
(© John Payne)

Figure 7.15 The sharp edge of Black Hill (Cat's Back, Fig.2.8) is in the foreground with the head of the Olchon Valley and the main Old Red Sandstone escarpment beyond. Hay Bluff is to the north (right) with the Wye valley behind it. (© Derek Foxton Archive)

Figure 7.16 Cross-bedded Brownstones Formation in former river cliff at Wilton Road, Ross-on-Wye (SO 596 240). (© John Stocks)

The Brownstones Formation

East and south of the Black Mountains, in the Forest of Dean for example, the Senni Beds are not distinguishable and all the uppermost beds of the Lower Old Red Sandstone (Fig. 7.3) are allocated to the Brownstones Formation. This Formation is 1,200m thick and is constructed almost exclusively of pebbly, cross-bedded, red-brown sandstones. It has provided the main building stone for the market town of Ross-on-Wye and is well seen in the nave of St Mary's Church (SO 598 240) where it is used extensively in both the 18th- and 19th-century renovation work.

Walking past the church through a large gateway to the cemetery, a superb view down the flood plain of the River Wye opens up. This site, known as the Prospect, sits on top of former river cliffs constructed of Brownstones which are best examined on the nearby Wilton Road close to the Royal Hotel (Fig. 7.16). Elsewhere, the Brownstones show excellent, well-documented sections along the M50 and A449, some of which have now been obscured by netting to prevent rock falls, and along the A40 near Pencraig (SO 5662 2224).[19]

Overall, the Brownstones become more arenaceous in character in their topmost beds where sandstones and thinner intraformational conglomerates dominate the sequence (Fig. 7.17). The conglomerates are rich in mudstone clasts incorporated by braided streams flowing over former mudflats together with a wide variety of exotic cobbles and pebbles.

On being used as building stones, the small, flattened mud flakes are the first to weather out, leaving distinct hollows in the rock surface. These may be seen, for instance, in Ross-on-Wye at SO 5980 2430 and in the nearby old river cliffs (SO 5955 2389).

Analysis of the rudaceous Brownstone sediments has identified clasts such as vein quartz, jasper, cataclasite, flow-banded and sometimes porphyritic rhyolite, fine-grained tuffs with poorly preserved graptolites, and a variety of sedimentary sandstones whose fossil contents suggest a Silurian age.[20]

The range of rock types is similar to that observed in the chronologically equivalent Woodbank Series in south Shropshire, although the latter horizon contains mid-Silurian Much Wenlock Limestone clasts which are absent in the Brownstones. Allied with palaeocurrent data, this suggests that the source area of the Herefordshire conglomerates lay in central and north Wales rather than the north-west English Midlands. The incorporation of Ordovician lavas and tuffs and Silurian sandstones together with a smaller proportion of Precambrian cataclasites and jaspers is evidence of

Figure 7.17 Intraformational conglomerate from the Brownstones Formation at Harewood End Quarry (SO 526 276) showing mud clasts and small fragments of calcrete. (© John Stocks)

considerable erosion of the Caledonian mountains by the end of the uppermost Lower Devonian.

The upstream areas of the alluvial plains in Lower Devonian times were characterised by a low cover of primitive plants, insufficient to retain water, with the result that the wet-dry monsoonal cycle produced alternating short-duration, high-discharge floods followed by a rapid drying out and desiccation. Overall, a landscape of primitive but sparse vegetation would lead to broad, shallow, braided drainage channels lacking cohesive retaining banks.[21] Individual flood events can be identified from their conglomeratic deposits within the successions, such as the Brownstones, and are found to persist from the Brecon Beacons through to the Herefordshire-Forest of Dean sequences.

The Brownstones Formation underlies much of the south central uplands of Herefordshire within what was formerly known as Archenfield, an area now of dispersed settlement, unusually devoid of a market town (Fig. 1.2). The well-developed sandstones form the high ground of Garway Hill and Orcop Hill and are exploited for building in a newly extended quarry at Harewood End (SO 526 276). Further south, the 12th-century Goodrich Castle (SO 577 200) is built on a Wye river bluff of the Brownstones which has also been extensively quarried to construct the castle. Hence the castle appears to grow out of its own bedrock (Fig. 10.8).

The largely arenaceous and rudaceous nature of the upper layers of the Brownstones reflects an increasingly close source area for the sediment. This can be equated with a major uplift to the north causing the erosion of the existing Lower Old Red Sandstone and the transport of the resulting sediment into areas further south. As the uplift continued the deformation front moved southwards across the Welsh Borders as did the source of the various major river systems. Herefordshire became an erosional area rather than a depositional zone and thus, at the top of the Lower Old Red Sandstone, there is a well-defined unconformity. This uplift and deformation has been recognised as the Acadian earth movements, which are now seen as a late phase of the Caledonian orogeny (late Silurian). During this time of non-deposition within the Middle Old Red Sandstone, there was gentle folding and erosion throughout the Anglo-Welsh Basin and later horizons of the Upper Old Red Sandstone, beginning with the distinctive Huntsham Hill Conglomerate, were laid down unconformably on the newly created land surface.

The Huntsham Hill Conglomerate Formation (Quartz Conglomerate)

The Upper ORS rock formations occur at the southern periphery of Herefordshire, where it merges with the uplands of the Forest of Dean. At the base of the Upper ORS, the Huntsham Hill Conglomerate Formation is 4 to 15m thick and outcrops around Chase Wood (SO 602 220) and Penyard Park (SO 619 224) within a large abandoned meander of the River Wye south and south-east of Ross-on-Wye. It clearly outcrops on Coppet Hill (SO 572 174) and, of course, at the type site on Huntsham Hill (SO 563 168) (Fig. 7.18). Further outcrops are found around the rims of the Howle Hill and Wigpool synclines with a particularly accessible outcrop at Euroclydon, on the Herefordshire border, north of Drybrook (SO 643 187).

Figure 7.18 A crag of Upper Devonian Huntsham Hill Conglomerate on Huntsham Hill, south Herefordshire. The inset shows a detail of the conglomerate (frame about six inches across).
(© Moira Jenkins)

The rock is made up of red-brown medium- to coarse-grained, pebbly sandstones with beds of quartz pebble conglomerate lying above the unconformity with the Lower ORS Brownstones. Rounded to well-rounded pebbles mainly of vein quartz (80%) but with subordinate quartzite, jasper and rhyolite lava are characteristic of this deposit, whose cross-bedding and pebble imbrication suggest vigorous, southward flowing braided streams. (Pebbles are said to be imbricated when they have been moved by the water flow to lie overlapping one another, an arrangement like that of roof tiles. This can show the flow direction.) The rock is particularly resistant to erosion but can be undermined by river and stream action on the underlying weaker rocks. This has given rise to the spectacular fallen boulders, the result of an ancient landslip (see Chapter 9), to be seen at the foot of the cliff beside the forestry track on Huntsham Hill below Yat rocks (SO 566 168). Perhaps the best exposures of the conglomerate lie just outside the county within the Forest of Dean, where crags such as the Suck Stone (SO 5421 1401) and Near Hearkening Rock (SO 5431 1401), both north of Staunton, are typical of this distinctive horizon.

Unlike the Lower ORS, the Upper ORS essentially represents a fining-upwards sequence with the largely arenaceous, rather than conglomeratic, Tintern Sandstone capping the succession on Chase Wood and Penyard Park Hill and forming outcrops around the Howle

Figure 7.19 Coningsby Hospital, Widemarsh Street, Hereford, built in about 1614 of coursed Old Red Sandstone blocks. (© Derek Foxton Archive)

Hill and Wigpool synclines. A good exposure of Tintern Sandstone is on Huntsham Hill (SO 5622 1648). The sandstones are mainly weakly cemented pale yellow-brown to pale greenish grey and are thought to have been laid down by southerly-flowing streams with the sandstones being channel-fill deposits and the finer mudstones and siltstones representing quieter floodplain environments. At the top of the formation, passage beds of thin fossiliferous marine shales and limestones indicate the transgression of the early Carboniferous seas over the ORS continent.

Old Red Sandstone as a building material

As often occurs, it may be easier to see good ORS rock samples in the walls of local buildings than *in situ*. Hereford offers good examples.

Herefordshire's Old Red Sandstone foundations are reflected in Hereford's medieval city walls dating from the 12th and 13th centuries. Coningsby Hospital, built in Widemarsh Street in 1614, is constructed of large blocks of coursed sandstone similar to that used in the city walls and obtained from the remains of Blackfriars monastery, a 13th-century house of the Knights Hospitaller (Fig. 7.19). The Norman cathedral, likewise, promotes its local geology with its magnificent columns in the nave built from light reddish coloured cross-bedded sandstone (Fig. 7.20). This is stone from the Freshwater West Formation.

Figure 7.20 A view across the River Wye in Hereford towards the Old Bridge and the cathedral, both mainly constructed of Old Red Sandstone. (© Derek Foxton Archive)

It was probably originally hewn from Capler Quarry (SO 590 325) near Fownhope and brought to the construction site overland by cart or up river on barges.

During later rebuilding, the west front of the cathedral used the more ornamental, pink and cream mottled Triassic Hollington Stone. Recent restoration work on the tower has drawn on the more resistant white- and red-banded middle Triassic sandstones of the Helsby Sandstone Formation sourced from quarries at Grinshill in Shropshire. Towards the Worcestershire border the use of Old Red Sandstone competes with the redder Triassic Helsby Sandstone (previously known as Bromsgrove Sandstone) in local parish churches. In one particular case, at Shelsley Beauchamp (SO 731 629) just outside the county on the fringes of the Abberley Hills, the earlier tower is constructed of Triassic Helsby Sandstone while the rebuilt Victorian nave is of more local Old Red Sandstone. Interestingly, this is the complete reverse of what is seen at Hereford Cathedral with Devonian replacing the softer Triassic building stone.

Devonian Old Red Sandstone was also much prized by the Herefordshire School of Sculpture, active in the second quarter of the 12th century, whose decorated corbels, voussoirs and tympani delicately carved from the red sandstones are such a feature of the Romanesque church at Kilpeck. Further examples of this work can still be seen in the Shobdon Arches, albeit somewhat weathered, and in Leominster Priory.[22] Unfortunately, the source of the fine-grained Old Red Sandstone used to such good effect at Kilpeck is not known.

Calcareous Tufa

Groundwater percolating through the Freshwater West Formation in north-east Herefordshire and parts of neighbouring Worcestershire has given rise within the last 10,000 years to another useful building stone, namely calcareous tufa. This is not part of the Old Red Sandstone but many occurrences locally are derived from the lime which cements many of the ORS rocks, notably at Southstone Rock (SO 7085 3955) in the Teme valley. Herefordshire occurrences of tufa are described in Chapter 9.

The end-Devonian mass extinction

Finally, we need to place our Siluro-Devonian Old Red Sandstone continent into a broader global context. Recent research has identified the third most important mass extinction event[23] as occurring within the late Devonian with 21% of marine families becoming extinct. This compares with the well-documented late Cretaceous 'Death of the Dinosaurs' event where only 15% of marine families perished.[24] Maximum extinction seems to have occurred in the latter half of the Frasnian (a 10My stage within the late Devonian at around 377Ma)[25] but the whole event can be seen as a broad extinction event with several smaller extinction episodes. ('Frasnian' and 'Famennian' are the names of subdivisions of the Devonian Period.)

The amalgamation of continental fragments, with the concomitant reduction in the length of available coastlines and their related continental shelves, is one of the main causes of a marked reduction in global biodiversity. By the late Devonian, within the time interval 383 to 359Ma, the global palaeogeography of only three relatively close major continents was already producing this expected reduction in diversity.

The lower diversity within the late Devonian world made it very vulnerable to any catastrophic event especially if that event directly affected one or two of the main three continents. With no major period of global volcanism identified, attention has turned to possible impacts by small asteroids especially as the Frasnian / Famennian interval is marked by a distinct iridium anomaly. This is the expected consequence of such an event. An asteroid large enough to cause the effects noted on the Devonian faunas would generate an impact crater at least 150km in diameter. Two candidates, although both much smaller in diameter, are in the frame. The largest of these is the 52km diameter Siljan crater in Sweden dated at 368Ma which equates well with the Frasnian / Famennian boundary at 367Ma. Another possibility is the Charlevoix crater in Quebec, Canada but its diameter of 46km and the fact that the 360Ma date has a published uncertainty of 25My makes this impact a less likely 'harbinger of doom'.[26]

Certainly, evidence of a huge loss in biodiversity is not disputed and the sharp changes in carbon isotope ratios recorded at the end of the Frasnian again point to a global catastrophe at this time.[27] Unfortunately, Herefordshire with its paucity of both floral and faunal remains within our Old Red Sandstone sequences is unable to shed much light on this event. Since it occurred at the very end of the Middle ORS just prior and probably during the deposition of the Huntsham Hill Conglomerate, a distinct lack of local rock horizons of this age is not helpful.[28]

The end of the Devonian was also marked by a global warming event and the retreat of the Gondwana ice sheets located at the South Pole. This initiated a sea level rise and a major marine transgression across the former Old Red Sandstone continent which, combined with northward drift towards the equator, led to the very different environments of the Carboniferous – a new chapter in our story of Herefordshire.

Further reading

For a recent appreciation of how the Old Red Sandstone underlies the major landscape features of Herefordshire, especially those of the Wye Valley, a recommended book is *Landscape Origins of the Wye Valley – Holme Lacy to Bridstow* edited by Heather Hurley (2008) and published by Logaston Press.

The guides from the Earth Heritage Trust on *Hereford Cathedral*, *Goodrich Castle*, *Queenswood & Bodenham* and *Ross-on-Wye* are particularly appropriate for studying the Old Red Sandstone. (See pages *xix-xx* 'EHT and BGS Publications'.)

An excellent guide, characterised by its useful field sketches, for the southern borders of the county is *Geology Explained in the Forest of Dean and the Wye Valley* by William Dreghorn (1968) and published by David and Charles. Although now out of print, Fineleaf Editions have produced a 3rd revised edition (2005) in a handy smaller format. A search on-line or in secondhand bookshops may turn up one of the various more recent editions of this well-known work.

Also see the Geological Conservation Review, vol. 31 (2005) *The Old Red Sandstone of Great Britain*, Barclay, W.J. *et al.* and Vol. 16 (1999) F*ossil Fishes of Great Britain*, Dineley, D. & Metcalf, S.

8 The Carboniferous, Permian and Triassic Periods

Herefordshire possesses few outcrops of rocks of Carboniferous or Permian age and these are confined to the extreme south and east of the county, in the Ross and Ledbury districts, where they form areas of distinctive landscape. They do, however, provide considerable information about the likely nature of the rest of the county area during these periods, since it is probable that at least some parts of the county were once covered by deposits of this age, now removed by erosion. Triassic rocks scarcely appear in the county.

The dawn of the Carboniferous Period in 'Herefordshire' saw a low-lying coastal plain to the south of the Old Red Sandstone continent and bathed in tropical heat. Figure 8.1 shows a possible interpretation of the geography of 'Herefordshire' at this time. The famous coal forests of the Carboniferous did not yet exist. The diagram shows the Early Carboniferous (355Ma) maximum transgression of the sea northwards onto the Wales-Brabant island (St

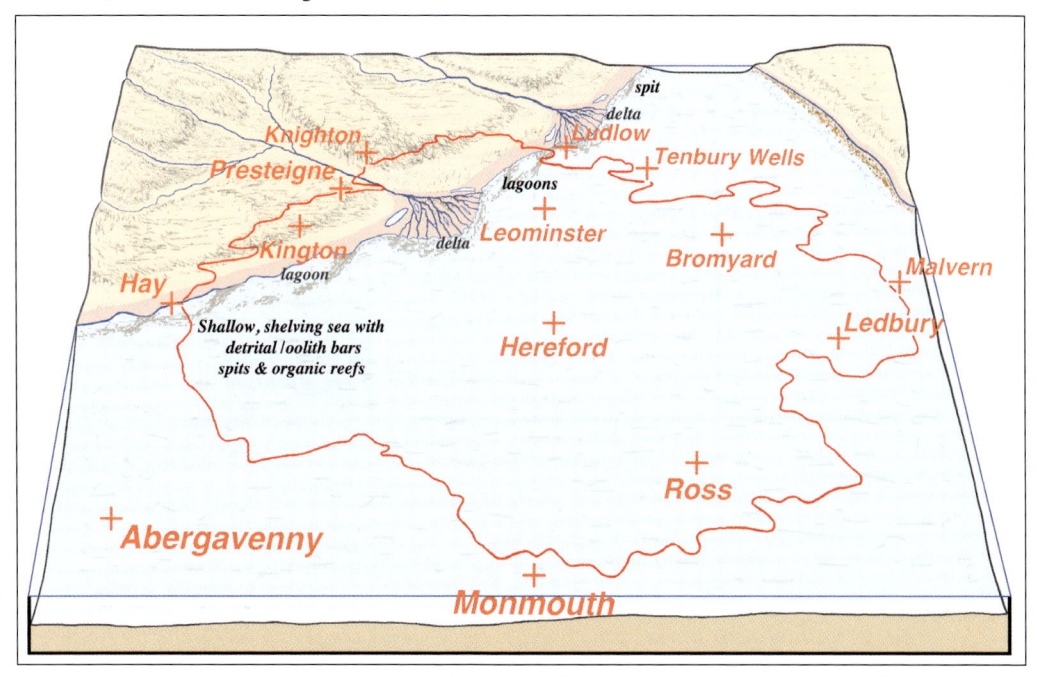

Figure 8.1 A suggested palaeogeographic view of Herefordshire in the early Carboniferous. (Drawn by Gerry Calderbank from a conceptual sketch by Dave Green)

George's Land), when limestones were being deposited in the area of Titterstone Clee, near Ludlow. The presence of biosparites, oolites, dolomites and calcretes (in the northern part of the south Wales coalfield) suggests a very shallow shoreline, evaporating, hypersaline lagoons and low coastal plains behind protective oolith or shell-detrital barriers, similar to the Trucial Coast of the Persian Gulf today. (Note that the positions shown for these in Figure 8.1 are entirely conjectural. The evidence, in the form of rocks of that age, was long since eroded away.) Sandstones were deposited by ephemeral rivers swollen by desert rainstorms. The palaeolatitude was between 10° and 20° south of the equator, so the area was possibly within the desert belt. As to relief, presumably the effects of the (mid-Devonian) Acadian orogeny had been removed by erosion, and the time of this picture is before the mid-Carboniferous folding (340 to 320Ma) (e.g. Forest of Dean and, possibly, Woolhope and Malvern). The later end-Carboniferous movements (320 to 300Ma) had a markedly smaller effect in the Forest. There is much less evidence for forests at this time than, for example, in Scotland, nearer the equator and in a wetter climate.

Our section of the Earth's continental crust had drifted northwards and lay virtually on the equator by 350Ma. To the north lay the remnants of the Caledonian mountain chain, levelled by 20 or 30 million years of erosion, by now producing only finer-grained

Carboniferous	Mushet 1824 [1]		Trotter 1942 [2]		Modern (BGS) 2007		
	COAL MEASURES	No. 2 (Coleford High Delf) (Trenchard)	UPPER COAL MEASURES	Pennant Group	WARWICKSHIRE GROUP	Pennant Sandstone Formation	Pennsylvanian
				Trenchard Group		Trenchard Formation	
			Unconformity		Unconformity (30 million years missing)		
		No. 1 (Millstone Grit)	CARBONIFEROUS LIMESTONE	Drybrook Sandstone	PEMBROKE LIMESTONE GROUP	Cromhall Sandstone Formation	Mississippian
	MOUNTAIN LIMESTONE	Upper Limestone Shale		Whitehead Limestone		Llanelly Formation	
		Limestone		Crease Limestone		Gully Oolite	
				Lower Dolomite		Barry Harbour Limestone Fm	
		Lower Limestone Shale		Lower Limestone Shale	AVON GROUP		

Figure 8.2 A table to illustrate the development of the terminology for the sequence of local rock units of the Carboniferous. Note that the unconformity in the middle represents a duration of half of the Mississippian and two thirds of the Pennsylvanian. (The Mississippian and the Pennsylvanian respectively correspond roughly to the Lower Carboniferous and the Upper Carboniferous.) This period of missing rocks lasted some 30 million years during which there was a major phase of folding followed by uplift and substantial erosion.
This book employs the modern terminology of the British Geological Survey (BGS).[3]

material. The nearest part of these remnants ran, in terms of today's geography, from southeast Ireland, across St George's Channel, through mid-Wales, the Midlands and on into Belgium; and is known as St George's Land (or the Wales-Brabant Massif). To the south lay the sea – a shallow, island-dotted tropical paradise with, further south, the deeper waters in the narrowing remnants of the Rheic Ocean. On the other side of this lay what is now Galicia in northern Spain, forming part of an Avalonia-like microcontinent ('Iberia-Armorica'). This comprised Iberia, Brittany and central France. (It is not shown in Figure 3.11, which shows the situation earlier, during the Devonian). Beyond Iberia-Armorica lay another stretch of ocean and then the supercontinent of Gondwana itself.

Between these continental blocks was emerging the developing Variscan mountain chain, essentially being produced by the collision of Gondwana and its outlying terranes with the supercontinent of Laurussia, including what is now Herefordshire. The continuing progression of this collision led to two enormous upheavals of the once flat-lying layers of sediments laid down over the previous 240 million years.

These ancient tectonic developments are hugely important for the current landscape. Despite the presence locally in today's land surface of few rocks from this time, the movements led to the way that present outcrops of rocks of different age and resistance to erosion come to the surface and thus to today's landscape eroded into them.

The various tectonic phases are described in this chapter but, first, a description is given of the rocks of this period which are to be found in Herefordshire, what they can tell us about the environment in which they were formed and their effect on the landscape. (The recently adopted current names[4] are followed by the previous names in brackets. The earlier names were the ones used in much of the literature on the local Carboniferous rocks. The correlations between the names are shown in Figure 8.2.)

The Avon Group (formerly Lower Limestone Shale)

This, the lowest rock unit of the Carboniferous sequence, is composed of many hundreds of beds. Outcrops are seen along the upper slopes of the Doward and Huntsham Hill, near Symonds Yat. It forms a plateau top on both sides of the Wye running from Lydbrook to Kerne Bridge, i.e. Courtfield, Welsh Bicknor and most of Howle Hill. Despite its earlier name, this formation is locally comparatively short of shale (thinly laminated hardened mud). Instead, it is composed mainly of grey limestone, often of the oolitic variety (composed of millions of tiny inorganic spheres of calcium carbonate, best seen with a hand lens on partly weathered surfaces) (see Box 8.1). Consequently, it is relatively resistant to erosion and forms upland areas. Its time of deposition is represented in Figure 8.1.

The limestone has been extensively quarried, mostly for lime burning in kilns to produce quicklime for mortar, especially on Howle Hill and Bishopswood, for example at Causeway Quarry, Howle Hill (SO 604 203) and in part-natural exposures on both sides of the Courtfield meander neck (SO 59 18). There is a fine example of a repaired limekiln on the lower slopes of Little Doward at SO 543 155, just inside the deer park boundary wall.

Occasionally, fossil brachiopods and solitary corals are found in the limestone. A new species of crustacean was found some years ago in a quarry on Great Doward.[5] This has been named *Schramocaris gilljonesorum*.

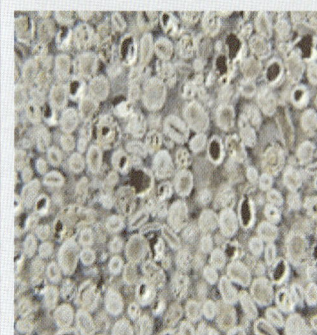

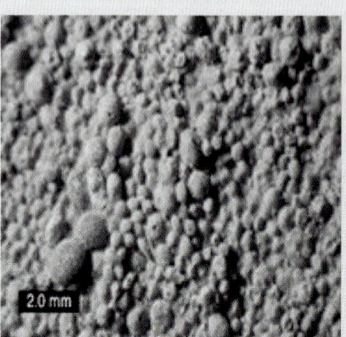

> Box 8.1 **Oolitic Limestones**
>
> Limestones today are being formed in warm tropical sea water. The water is very clear, with no mud or sand to mask the slowly deposited lime. The warmth helps lime to precipitate from its dissolved state, being itself dependent on the amount of dissolved CO_2, which is less soluble in warm water.
>
> Ooliths (Fig. 8.3) are small (1 to 2mm) spheres composed of calcium carbonate precipitated in concentric layers onto the surfaces of tiny objects, such as sand grains or shell fragments, by the gentle to-ing and fro-ing of water on the sea floor caused by wave action. Using the geologists' Principle of Uniformitarianism, the presence of such rocks indicates deposition in extremely shallow water, probably less than three metres and probably therefore close to shore.
>
> *Figure 8.3 Oolitic limestone in the Avon Group. (© Dave Green)*
>
> Polished section through oolitic limestone — Surface of oolitic limestone — Oolitic limestone from near King Arthur's Cave

The Pembroke Limestone Group (formerly Carboniferous Limestone)

Above the Avon Group rocks lie those of the Pembroke Limestone sequence. These outcrop on the Doward and extend eastward along the main escarpment at English Bicknor, above Ruardean, Hope Mansell and Lea, but not on Howle Hill or the Courtfield plateau. Massive crystalline dolomite limestones of the Barry Harbour Limestone (Lower Dolomite) and Gully Oolite (Crease Limestone) divisions of the Pembroke Limestone form the spectacular cliffs lining the Wye Gorge, such as Coldwell Rocks (where the well-known and much spied-upon peregrine falcons nest each year), on the right (west) bank of the Wye south of Symonds Yat, and Seven Sisters Rocks. The Gully Oolite is the main repository of the red goethite iron ore and has been quarried, in the search for the metal, both on the surface and underground right along these outcrops. There are some natural caves too, such as King Arthur's Cave (SO 5458 1558) (Fig. 2.25) and Merlin's Cave (SO 556 153). The latter has been disturbed by the iron miners but both have yielded late Ice Age faunas such as mammoth, lemming and giant deer.

In upward sequence the units forming the local variant of the Pembroke Limestone Group are:

Barry Harbour Limestone Formation (formerly Lower Dolomite Formation) (Fig. 8.4). This is a relatively fine-grained dolomitic limestone. It is composed of calcium magnesium

Figure 8.4 Barry Harbour Limestone (Lower Dolomite).
(Left photo) Note the dense sugary fine crystalline nature of this much-quarried rock, and the typical purplish grey colour.
(Right photo) A thin section, cut to 30 microns thickness to allow light to pass through and enable microscope viewing of the rhombic crystals of Barry Harbour Limestone.
(© Dave Green)

carbonate, rather than the calcium carbonate of most limestones. Two simple tests can determine which you have. Limestone (hardness 3) will be scratched when the edge of a copper coin (hardness 3.5) is drawn across it, whereas dolomite (hardness 3.75) will not. Also, limestone will fizz when cold dilute hydrochloric acid is added to it whereas dolomite will not.

This is currently the most-quarried rock in the Forest of Dean because, although it is relatively soft and easy to crush into pieces, it has great load-bearing strength when used as an aggregate in concrete and for road foundations. Good examples of exposures of this massive rock are the large quarry at Drybrook and the quarried cliff running north-west from Symonds Yat on the opposite side of the river to the Saracens Head public house.

Figure 8.5 Thin section of dolomitised Gully Oolite (Crease Limestone). The dolomite crystals are often 2 to 3mm across. Note their rhombic form and that of the cleavage planes traversing them. (© Dave Green)

Gully Oolite Formation (formerly Crease Limestone Formation). One of the most distinctive rocks in the Forest of Dean, this is something of a misnomer in that neighbourhood, since there it is not an oolite but is a coarse-grained dolomite, usually a straw colour, but often stained pink, and without a trace of ooliths! (See Fig. 8.5). It is composed of a mass of coarse (1 to 3mm) intergrown rhomb-shaped dolomite crystals whose planes of cleavage sparkle on freshly broken surfaces. The reason for the coarse grain size, in contrast to the Barry Harbour Limestone, is that the Gully Oolite was originally an oolitic limestone deposited in agitated water, whereas the Barry Harbour Limestone was a fine-grained limestone deposited in quiet water.

| Box 8.2 | **Dolomitisation** |

Most of the Carboniferous limestones in the Forest of Dean have been transformed, just after deposition of the sediment, by the process of dolomitisation – the addition of magnesium, probably by circulating water which replaced some of the original calcium, causing recrystallisation and the destruction of many of the original features of the rock such as ooliths and fossils. Calcium carbonate ($CaCO_3$) is altered to calcium magnesium carbonate, $CaMg(CO_3)_2$. This is thought to be especially associated with nearness to coastlines where there is an interaction between circulating fresh and salt water in the ground. It is noteworthy that the same formations become less and less dolomitised as they are traced southwards towards Bristol and the Mendips, away from the shore of ancient St George's Land.

The **Llanelly Formation** (formerly Whitehead Limestone Formation) (Fig. 8.6), also dolomitised, is by contrast extremely fine-grained and looks like grey flint in hand specimen. It represents the lime-rich sediment slowly settling out of suspension from the warm waters of a lagoon. From time to time, sticky algal mats, trapping thin layers of lime, grew up into the water and on the shoreline to form dome-like stromatolite colonies (similar to Shark Bay in north-west Australia today). This formation has been rarely quarried, being brittle, lacking strength, containing numerous clay-rich beds, and with a tendency to

Figure 8.6 Lord's Wood Quarry (SO 548 155) in the Llanelly Formation and the underlying Gully Oolite Formation. Note the gentle dip to the right here on the flat-lying western limb of the Ross Syncline. (© Dave Green)

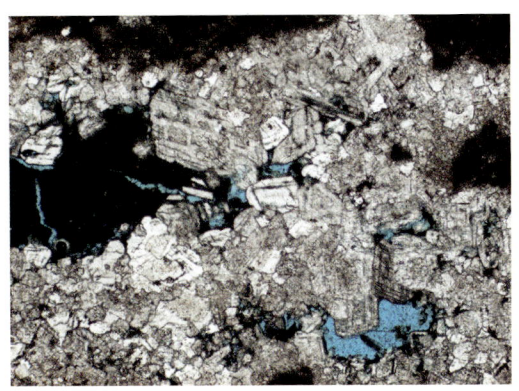

Figure 8.7 A stained thin section of Llanelly Formation limestone showing clearly the rhombic crystals of dolomite and their rhombic planes of weakness (cleavage planes). This thin section is at a greater magnification than Figs. 8.4 or 8.5 because the rock is so fine-grained. The blue areas show where the blue stain has entered porous crystals. Some appears to be in fractures and some in pores developed because of shrinkage on dolomitisation. The dark areas probably show fine-grained clay minerals. (© Dave Green)

shatter into small pieces rather than form the large blocks needed for building purposes. An exception is Lord's Wood Quarry (SO 548 155; near King Arthur's Cave; Fig. 8.6) but the main rock quarried here is Gully Oolite. A walk around the area of smallholdings to the north of Lord's Wood will reveal numerous small outcrops of white-weathering Whitehead Limestone both in the beds and in the banks of the numerous tracks that cross this area.

The **Cromhall Sandstone Formation** (formerly Drybrook Sandstone Formation) forms the top-most member of the Carboniferous Limestone sequence. It is a multi-coloured (red, brown, yellow and white), poorly cemented sandstone, very friable and often quarried for building sand with the minimum of crushing. Towards its base it often contains white quartz pebbles, much rounder and often smaller than those seen in the older Huntsham Hill Conglomerate (Quartz Conglomerate). This was probably a shoreline or coastal river deposit. It is found on top of the Doward, where it creates patches of relatively acid sandy soil, in contrast to the thin, stony clay-rich alkaline soils developed on the underlying limestones. Hence the lack of farming on Lord's Wood, where it is possible to find loose blocks of the sandstone, especially in the roots of trees blown over by the wind.

It is not clear how far these deposits may once have extended to the north of their present outcrop. They were probably present and have been eroded from where the southerly dip (perhaps imposed during the Asturian orogenic phase [320 to 300Ma]) elevated them. As they are shallower water deposits than those to the south, for example in the Mendip Hills, it is probable that they did not extend as far as Hereford. Hereford lay on St George's Land, to which there may well have been a southern faulted boundary, now, as then, concealed below Old Red Sandstone.

The mid-Carboniferous earth movements (Sudetic phase of the Variscan Orogeny)

In south Wales and the north of England, the rocks of the Pembroke Limestone Group (Carboniferous Limestone) were followed by the deposition of the deltaic deposits of the Millstone Grit, forming many of the Gritstone moorlands, such as Kinder Scout, or nearer home, the plateau of the Black Mountain, south of Llandeilo. In our district, this division of the Carboniferous (the Namurian) is completely missing, as are the lower and middle divisions of the succeeding Coal Measures, laid down on the top of the delta in swamps with distributary channels meandering through. Why should this be so? Study of the geological map of the Symonds Yat to Howle Hill area shows that something extraordinary has happened. Just south of Yat Rock the Trenchard Formation (Upper Coal Measures) rests on the Cromhall Sandstone (Drybrook Sandstone) at 140 metres above sea level, whereas on Howle Hill the Trenchard Formation rests on the Avon Group (Lower Limestone Shale) at the same height (Fig. 8.8). This represents a large unconformity which includes the already large mid-Carboniferous one indicated in Figure 8.2 plus the absence of the Pembroke Limestone Group (Carboniferous Limestone). The underlying rocks were compressed and folded (probably by the collision of Brittany and Spain to the south), uplifted well above sea level, and then were eroded, removing all the Pembroke Limestone sequence from Howle Hill but not from Wigpool and Symonds Yat, where the Cromhall Sandstone was preserved in the troughs of synclines (or downfolds).

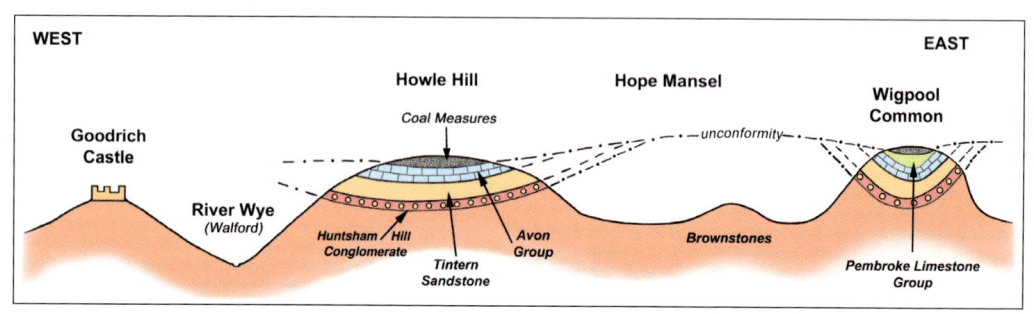

Figure 8.8 A cross-section to show the effects of two periods of earth movement: the limestone of the Pembroke Limestone Group (Carboniferous Limestone) and earlier rocks, were folded during the mid-Carboniferous, and then eroded. During the Upper Carboniferous, the Trenchard Formation (Warwickshire Group) was deposited on the planed-off surface (the surface of unconformity), and then all the rocks were folded again at the end of the period, though more gently. Erosion was deeper on what is now Howle Hill, removing most of the Pembroke Limestone, compared to Wigpool.
(Drawn by Gerry Calderbank from a conceptual sketch by Dave Green)

The Warwickshire Group (Upper Coal Measures)

Subsidence of the area, an essential condition for the accumulation of sediments of any thickness, resumed towards the end of Carboniferous times. Swampy conditions prevailed, indicated by the presence of coal seams, which were once thick beds of peat, prevented from breaking down by waterlogging (which reduced the access to oxygen), and mudstones deposited from virtually still water. These conditions were broken only by the migration of sand-filled channels through the swamps. This sequence of grey mudstones and khaki-coloured sandstones, with the occasional coal seam, constitutes the Trenchard Formation. It forms the top of Howle Hill and there are a few remaining patches on the top of the Welsh Bicknor (or Courtfield) peninsula (SO 59 18). Howle Hill, is also known as 'Coal Hill' and, at Great Howle Farm, was the site of small-scale open cast coal mining between 1972 and 1977. It is virtually certain that this Trenchard Formation once extended much further north across Herefordshire, and was itself overlain by the much thicker Pennant and Supra-Pennant Formations, still preserved at Hillersland, Ruardean and Harrow Hill to the south in the Forest of Dean. Remnants of these beds are found preserved in the Malvern Fault belt, for example in the tiny Newent coalfield. To the north, in the West Midlands and Shropshire, the Upper Coal Measures were deposited on top of parts of St George's Land. They have been mostly removed by subsequent erosion, probably during the Permian Period.

The end-Carboniferous earth movements (the Variscan Orogeny)
The end of the Carboniferous Period, about 300 million years ago, saw the final collision of the supercontinents of Gondwana and Laurussia to form Pangaea. Virtually the whole of the area we now call Britain was uplifted to form a small central part of this vast landmass, which comprised most of the land on the planet. The main collision occurred to the south, for example between Brittany and Spain, creating a Himalayan-style mountain chain stretching from west to east, from the southern Appalachians to Bohemia, of which Devon and Cornwall formed part of the northern foothills. (The opening of the Atlantic Ocean was still far in the future.) This mountain-building episode is known as the Variscan Orogeny. Folds and great breaks in the chain, known as thrust faults, moved masses of folded rock northwards, rather like giant east-west trending waves moving up a beach. These 'waves' reached as far as a line through Bristol and running west-north-west through south Wales and Pembrokeshire and on to south-west Ireland; this line is known as the 'Variscan Front'.

Our area lay to the north of the Variscan Front but the event still produced major structural changes in the Herefordshire rocks, notably in the east of the county with the raising of the Malvern rocks to their present position and leading to the formation of today's Malvern Hills. (The role of the Variscan Orogeny in the formation of the Malvern Hills is described also in chapter 4.) We will discuss first, though, the relatively minor effects of the Variscan Orogeny in the remainder of the county.

Figure 8.9 shows the main structures of Variscan origin in the remainder of the county. These Variscan structures have had a major effect in controlling the development of the present

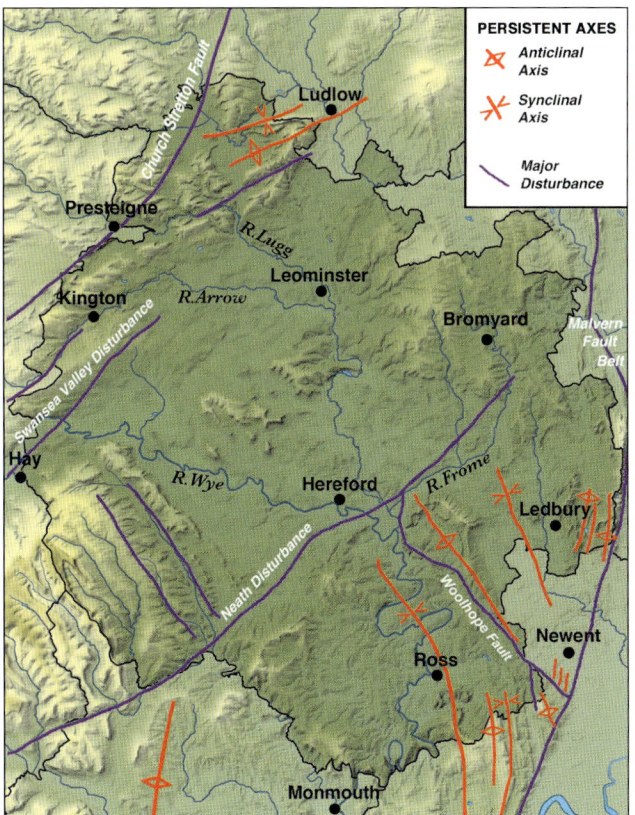

Figure 8.9 The major faults and folds of Herefordshire and their relationship to overall topography. (Drawn by Gerry Calderbank from a conceptual sketch by Dave Green)

landscape. The varied trend of the folds is due to the varying orientation of underlying basement faults, which moved to produce the folds, as described later. In turn, the orientation of uplands and lowlands, ridges and valleys, is due to the effect of folding and faulting through bringing layers of rocks of differing resistance to erosion to the surface. Mudstones, shales and siltstones are relatively weak and have eroded into low-lying flattish country whereas the more erosion resistant limestones, sandstones and conglomerates stand up proud as ridges and escarpments.

As illustrations of this, Figures 8.10, 8.11 and 8.12 show geology and scenery relationships looking north and south of Ross and the landscape features resulting from the differential erosion.

At Bishop's Frome (SO 662 483) lies the north-east end of a fault system known as the Neath Disturbance (Fig. 8.9). This has a north-east to south-west trend (known as a Caledonoid trend, i.e. Caledonian in direction but not in age) and separates Shucknall Hill (SO 588 430) from the main mass of Silurian rocks in the Woolhope Dome to the south-east. The Neath Disturbance was originally part of a family of such faults marking the south-east margin of the Welsh Basin during Lower Palaeozoic times. As a line of

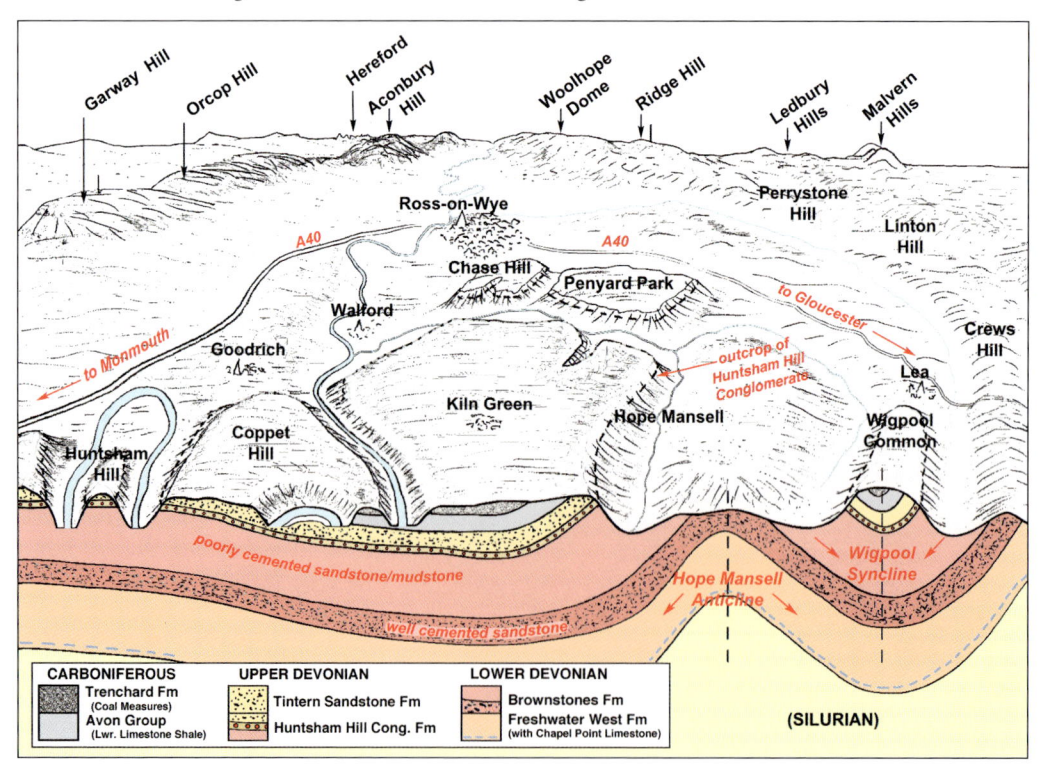

Figure 8.10 Geology and scenery looking north towards Ross, Hereford, the Malverns and the Woolhope Dome. This shows the northern end of the Carboniferous outcrops, the minor folds and the major Ross Syncline, picked out by the horseshoe-shaped ridge and hills formed by the well-cemented sandstones in the lowest parts of the Brownstones Formation. In the section, the Ross Syncline is shown underlying Kiln Green and Coppet Hill. The following Figure 8.11 is drawn looking south. (Drawn by Gerry Calderbank from a conceptual sketch by Dave Green)

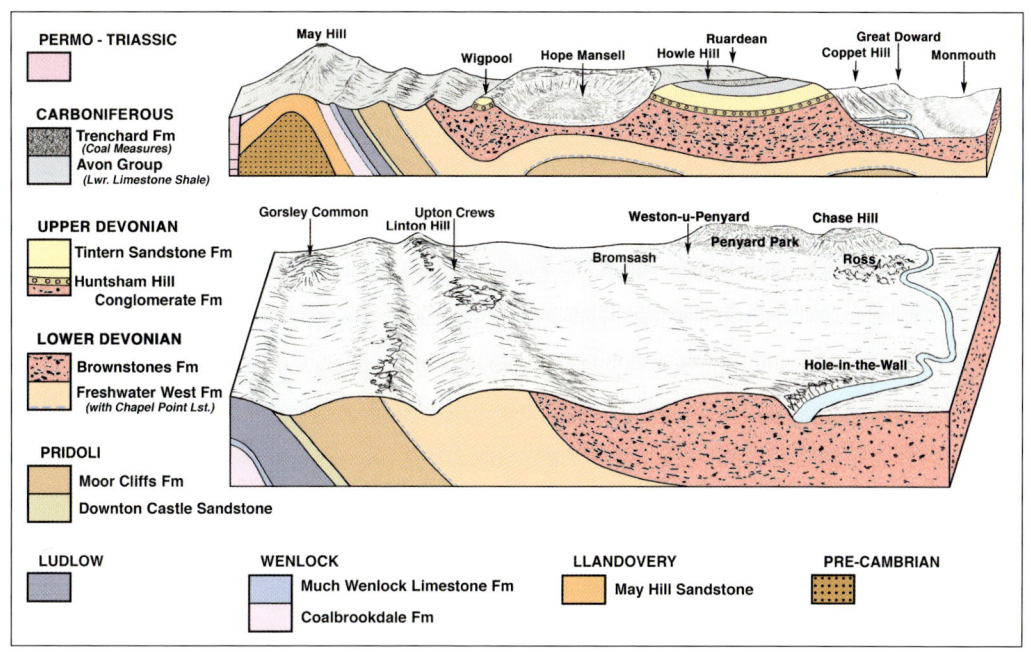

Figure 8.11 Geology and scenery looking south towards Ross and the Forest of Dean / May Hill. The effect of folding in bringing resistant and more easily weathered rocks repeatedly to the surface is seen, along with the difference in shape of landform due to the changing amount of dip. Overall, the rocks are dipping southwards, away from the observer, meaning that the younger units appear in that direction. The general structure is a broad asymmetrical syncline, steepening towards the Malvern Fault (at the left end of the upper diagram). Additional subsidiary folds are developed towards the south, notably the Hope Mansell anticline and the Wigpool syncline. (Drawn by Gerry Calderbank from a conceptual sketch by Dave Green)

Figure 8.12 The Upper Old Red Sandstone/Carboniferous Limestone escarpment viewed looking north-east from Coppet Hill (SO 578 184). The escarpment is broken by existing and former incised meanders of the Wye. The nearer bluff is Howle Hill, capped by the Avon Group (Lower Limestone Shale), with the Trenchard Formation (Coal Measures) forming the peak to the far right. The further bluff is Chase Wood, capped by Huntsham Hill Conglomerate (Quartz Conglomerate) and Tintern Sandstone. (© Dave Green)

weakness, it was reactivated during the Variscan Orogeny, fracturing and warping beds in its vicinity, and was the conduit up which magma rose at the end of the Carboniferous to form the Bartestree Dyke, exposed in the old quarry on Lowe's Hill (SO 566 405), near St Michael's Hospice. This dyke, which is about 13 metres wide and 200 metres long, is composed of dolerite, a hard, dark crystalline rock used especially for roadstone. Most has been quarried away, but the remains of the dyke are still visible at the south-western end of the old quarry. Here, its heat has baked hard the surrounding mudstone of the Moor Cliffs Formation (Raglan Mudstone) and bleached its colour. (A similar dyke and quarry may be seen at Brockhill near Shelsley Beauchamp [SO 7263 6380].) In the north-west of the county, the pre-existing faults of the Church Stretton group, including the north-east to south-west trending Swansea Valley Disturbance and the Leinthall Earls fault, were also re-awakened and are probably responsible for the formation of the Caledonoid Ludlow anticline (Wigmore Dome) and Downton syncline.

These were the numerous effects of the Variscan Orogeny in Herefordshire. However, the major local effect of the Variscan Orogeny was the compressive reactivation of the ancient Malvern Fault and its branches, including the Woolhope Fault with attendant over-turning of beds, the formation of both low angle thrusts and up to 900m of movement westwards up the compressive Woolhope Fault (Fig. 8.13). Both produced parallel 'ruck' folds, including the Woolhope Dome,[6] due to 'sticking' on these faults. The approximate form of the present ground surface along the line of the section is shown figuratively in the diagram; all the material above this line has been removed by subsequent erosion, much of it, no doubt, while the thrusting was in progress.

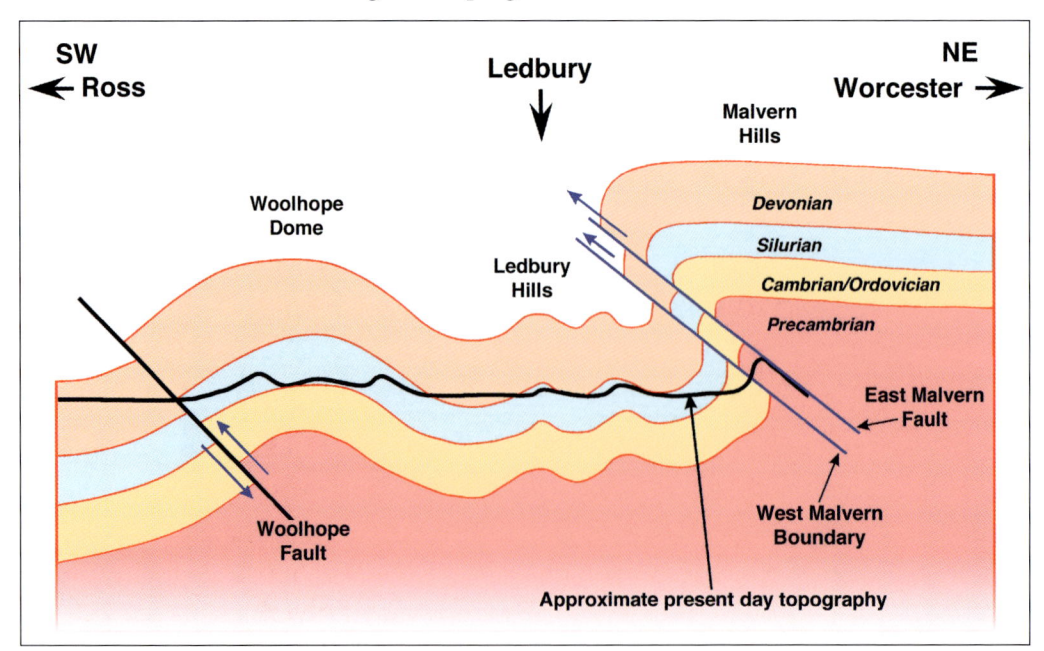

Figure 8.13 Sketch cross section showing the effect of the Variscan Orogeny on the east of the county (c.300Ma).
(Drawn by Gerry Calderbank from a sketch by John Payne)

As in the previous earth movements, the crust was moved violently but very slowly along pre-existing planes of weakness running through the rock, particularly the north-south trending Malvern Fault and the north-west to south-east trending Woolhope Fault which branches off from it at May Hill (Fig. 8.9). Rocks from the Precambrian to Carboniferous to the east of these dislocations were bodily pushed westwards, at first folding under the pressure then finally breaking and being thrust westwards up the fault planes (Fig. 8.13).[7]

The folds produced next to the faults were sharp and often steep but this effect quickly diminished to the west, producing gentle open folds such as the Ledbury syncline (between Ledbury and the Woolhope Dome), the Forest of Dean Coal Basin and its northward extension – the Ross-Hereford syncline. In our area the folds generally plunge southwards (i.e. the axes of the folds are tilted down to the south), meaning that younger beds usually outcrop to the south and older ones to the north.

By approximately 10 million years after the Variscan Orogeny, erosion had stripped off all the Lower and Upper Palaeozoic cover (a thickness of about 2.5 kilometres of rock!) from the uplifted area to the east of the Malvern Fault, exposing the underlying Precambrian basement as hills to the east of the present line of the Malverns (Fig. 8.14).[8]

At this time, the very end of the Carboniferous (the Stephanian stage), the landscape was one of rugged rocky hills, uplifted by the force of the continent-building collision. There may have been some similarities to today's landscape, in the sense of the positioning of bands of hard and soft rocks. The major exception to this was in what is now the Severn Vale. This, the upthrust side of the Malvern Fault belt, was probably much higher than the Welsh Borderland side even after the removal of the Palaeozoic rocks. (We know from a deep borehole drilled at Kempsey near Worcester that all the Cambrian, Ordovician, Silurian, Old Red Sandstone and Carboniferous rocks were eroded here before Permian sediments were deposited on top.)[9]

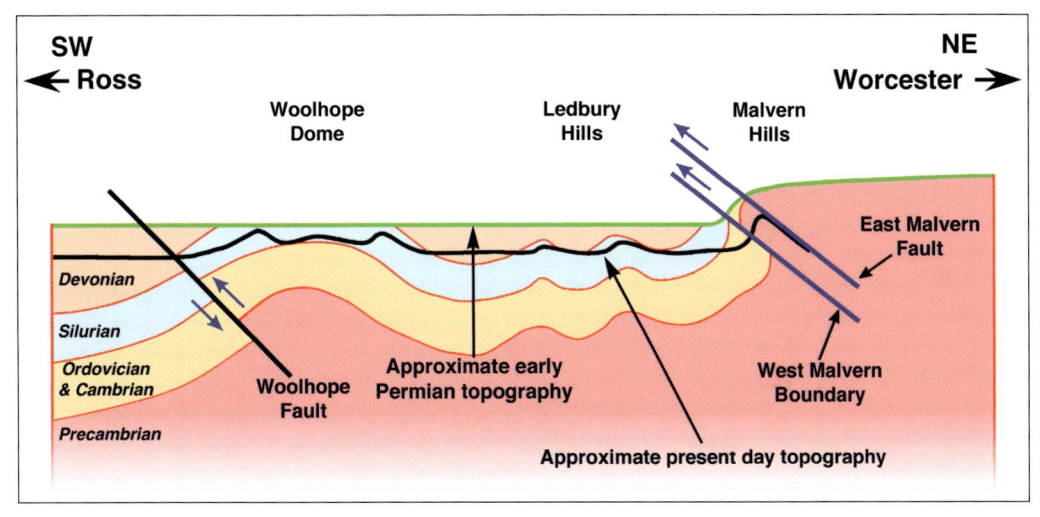

Figure 8.14 Diagram showing the effect of approximately ten million years of erosion following the Variscan Orogeny.
(Drawn by Gerry Calderbank from a sketch by John Payne)

The Permian Period

Almost as soon as they had formed, once the impetus of the Variscan collision had petered out, the mountains and hills started to collapse, partly through erosion, but mainly through tectonic forces. The previous compression was reversed and the Malvern Fault became an extensional structure (a normal fault wherein the upper side of the fault moves downwards under gravity). In particular, the eroded 'horst' (or upfaulted block) to the east of the Malvern Fault subsided to form a north-south trending rift valley (or 'graben') which became the 'sink' for most of the pebbles, sand and clay generated by erosion of the surrounding hills (Fig. 8.15). Today this forms the low-lying Severn Vale, containing up to 1km thickness of Permian desert sediments followed, above, by 2km from the Triassic Period and some Jurassic sediments on top, all deposited in the succeeding 100 million years (proved in the 1997 Kempsey borehole[10]).

For some unknown reason, the small, westernmost element of the Precambrian horst did not take part in the collapse into the rift valley. It still stands today as the Malvern ridge, a commanding feature in the scenery.

In Figure 8.15, the western boundary of the rocks of the Malvern Hills is shown as a fault. This boundary is little exposed and often appears as an unconformity between the Precambrian and the Palaeozoic rocks. It is thought likely that the boundary is in some locations a fault and that in other places a small fault exists within the adjacent Silurian rocks and the boundary shows at the surface as an unconformity.[11]

The Permian rocks are deeply buried below younger Triassic and Jurassic rocks of the Severn Vale but are exposed in the tract of country to the south of the Malvern and Ledbury Hills and to the north of May Hill. In this district, the beds are preserved between two of the faults making up the Malvern Fault Zone. (The fault zone is more complicated than the simple two-fault structure shown in Figs. 8.13 to 8.15.) They are gently inclined (or 'dip') towards the south, so that the older Permian beds are exposed in the north and the younger Triassic rocks in the Newent district of Gloucestershire to the south.

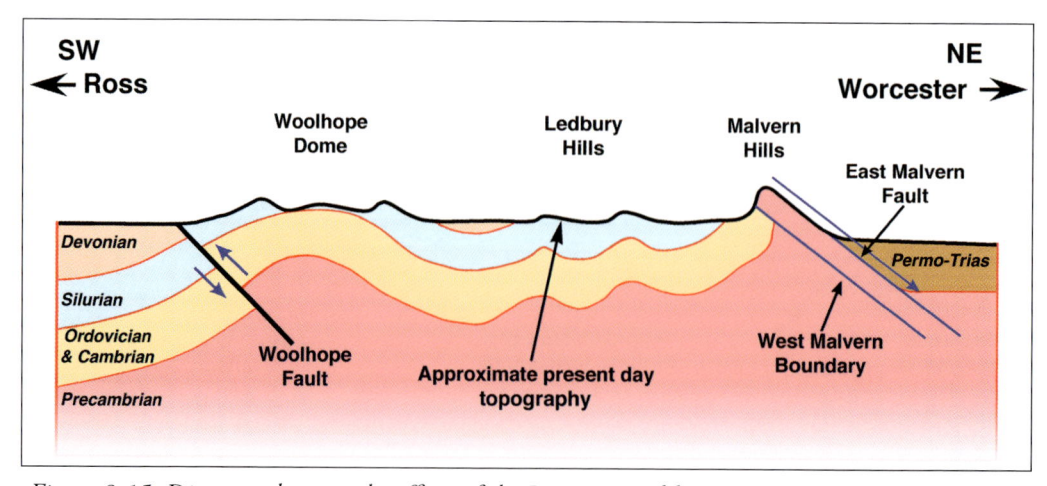

Figure 8.15 Diagram showing the effects of the Permian and later extensional movement on the Malvern Fault system to generate the chief elements of today's topography. (Drawn by Gerry Calderbank from a sketch by John Payne)

Lying directly on top of the eroded edges of the older folded rocks of Precambrian to Devonian age is the oldest division of the Permian, the **Haffield Breccia**. This rock is roughly exposed in the banks of the lane (around SO 749 348) leading towards Whiteleaved Oak from Bromsberrow but is displayed in all its glory in an old quarry in the grounds of Haffield House (SO 724 337). Here it has been tastefully incorporated into the design of the garden, as well as being the main building stone used in the construction of the house (Fig. 8.17).

The Breccia consists of sharp-edged (angular) fragments of the older rocks held in a muddy red matrix (Fig. 8.19). This arrangement suggests that the fragments were not moved very far (otherwise they would have been rounded by erosion during the movement) and that they were moved as a type of slurry (only fast currents of water could have

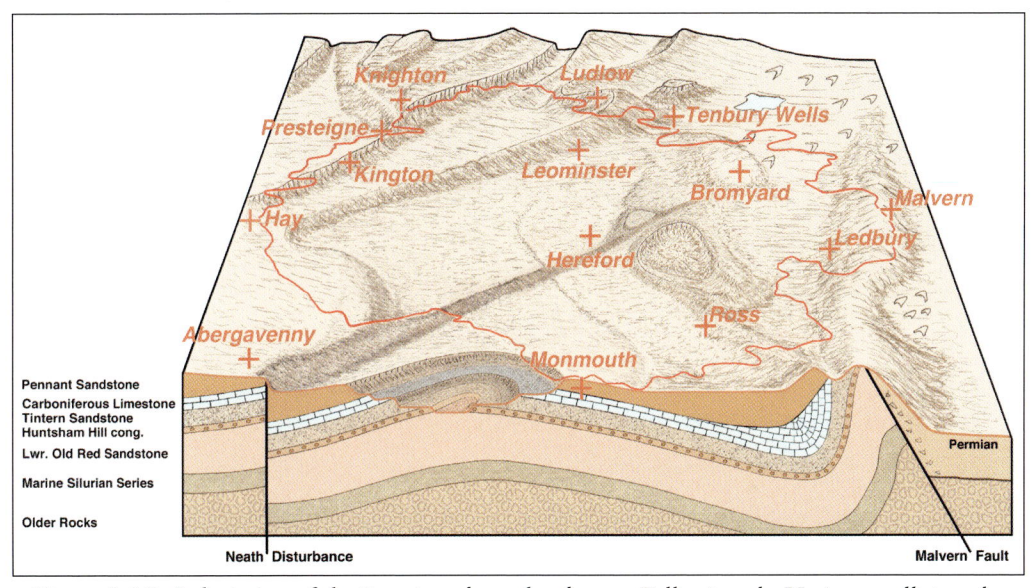

Figure 8.16 A depiction of the Permian desert landscape. Following the Variscan collision that formed the supercontinent of Pangaea, the layers of rock laid down during the previous 300 million years were disrupted. This occurred especially near to the fundamental fractures cutting through the Herefordshire crust: the Malvern Fault and its branches (such as the Woolhope Fault); and the Church Stretton Fault and its offshoots (such as the Leinthall Earls-Clee Hills Fault). As these faults were jostled by the collision to the south, the overlying rocks were buckled into folds, some blocks of crust being raised to form uplands. Being in a desert climate, wadis and canyons were eroded by flash floods. This and physical weathering due to temperature cycles and the growth of salt crystals, which caused stresses within the rocks, produced a landscape that was probably reminiscent of the south-western USA today. The collapse of the eastern side of the Malvern Fault soon after the collision (Fig. 8.15) led to the accumulation of large thicknesses of river- and wind-deposited sediments in what is now the Severn Vale. There has been no further major period of earth movement since the Variscan, so the Permian, Triassic and later strata which remain to the east of the Malverns all lie nearly horizontally except for a small dip to the south-east (see Chapter 9).
(Drawn by Gerry Calderbank from a conceptual sketch by Dave Green)

Figure 8.17 The quarry in Haffield Breccia at Haffield House (SO 724 337). This was the main source of stone used to construct the house. An unusual building stone, it is difficult to cut to the right shape around the irregular large rock fragments it contains. (© Dave Green)

Figure 8.18 The Haffield Breccia escarpment, viewed from the west near Brooms Green (SO 715 329). The low ground in front is developed on the poorly cemented and highly permeable Bridgnorth Sandstone, complete with outdoor pigs. Haffield House is the large white building in front of the trees. (© Dave Green)

Figure 8.19 A close up picture of the Haffield Breccia – an assortment of local rock fragments, including Malvernian granites and diorites. These fragments are angular in shape and are embedded in a muddy red matrix, suggesting a desert mudflow originating in the rocky hills newly uplifted by the Variscan Orogeny. (© Dave Green)

moved such large fragments of rock, but the fast currents would have washed away any fine mud). It is likely that the fragments were eroded from the raised area of Precambrian rock depicted in Figure 8.14.[12] As with the Old Red Sandstone, the red colour indicates the presence of the iron oxide mineral haematite, suggesting hot dry conditions. (You can turn yellow-brown iron oxide, on iron or in rocks, to red by heating, which drives off the water.) So, our interpretation of the landscape of the time is of a hot desert landscape with hills to the north of what is present-day Bromsberrow and from which issued flash floods and/or mudflows after occasional torrential rains (Fig. 8.16). The lack of vegetation in similar areas today means that water running off the hills rapidly soaks into the loose, dry material on the ground, quickly turning it into liquid mud which is capable of carrying large rock fragments. Palaeomagnetic readings on the Permian rocks, which contain small quantities of the mineral magnetite, suggest that the area lay at about 10° to 20° north of the equator at this time, about 270-290 million years ago.[13] The lack of accuracy in dating is because, firstly, we cannot directly date these particular sedimentary rocks radiometrically and, secondly, the Breccia contains no fossils which we could use to correlate their age with other outcrops such as those near Perm, at the foot of the Ural Mountains in Russia.

Above the Haffield Breccia lies the **Bridgnorth Sandstone**. This is well exposed just over the border in Gloucestershire in quarries and in the banks of lanes. It is present in the low ground to the south of Haffield House. It is a distinctive bright red sandstone whose grains are very poorly cemented together. Consequently, it breaks down easily into loose sand and absorbs water like a sponge – this makes very good ground for keeping pigs, especially in winter (Fig. 8.18)! There are scattered exposures of the sandstone in the banks of sunken lanes worn down into the rock by constant use, but the best place to see them is in Bromsberrow Lane (around SO 748 344, just outside Herefordshire), south of the outcrop of Haffield Breccia. Here you will appreciate the fine layers, or laminations, in which the sand accumulated and the way that the laminations cross one another at different angles (known as cross-bedding). If you have a hand lens, inspecting the sand grains will show them to be all the same size and so well rounded as to be polished. These are the classic signs of desert sand dunes, where wind-blown sand accumulates on the sheltered downwind side of the dune by avalanching down from the crest, building up thin layers on that slope. The

Figure 8.20 The Bridgnorth Sandstone exposed in Bromsberrow Lane (near SO 748 344). The lane here is sunk into the rock probably due to the erosive effect of cartwheels over the ages. The rock is poorly cemented and is hence very permeable. Although the beds are actually dipping very gently south, they appear to be dipping at different angles; this is due to cross-bedding produced by the migration of desert sand dunes and the avalanching of sand sheets down their steep fronts. (© John Stocks)

fact that most of the cross bedding is inclined towards the south (towards Bromsberrow) strongly suggests that the prevailing winds came from the north. This is another piece of evidence backing the idea that this country lay in the Trade Winds belt to the north of the equator, where the modern Sahara is situated (Fig. 8.20).

It is possible that the Bridgnorth Sandstone was deposited over much of Herefordshire but none now remains.

The Mesozoic Era
The Mesozoic Era (Fig. 3.1) comprises the Triassic Period, which is seen in and close to Herefordshire, and the Jurassic and Cretaceous Periods, which are not seen locally.

The Earth's most severe extinction event occurred at the end of the Permian Period, concluded the Palaeozoic era and ushered in the Mesozoic era. 96% of marine species became extinct together with 70% of terrestrial invertebrate species. The worldwide fossil record changed enormously. This event is not recorded locally however. Rocks of this age are missing from Herefordshire but are thickly developed in the neighbouring Severn Vale. There, we know through information from boreholes that the Permian Bridgnorth

Sandstone is overlain by a thick sequence of Triassic rocks including mudstones, sandstones and pebbly conglomerates, all laid down in the subsiding trough we call the Worcester Graben. These rocks demonstrate an extreme desert environment which allowed little life to exist so there is an almost complete absence of fossil remains.

To the south of Bromsberrow, just outside Herefordshire, there lies an interesting district of steep valleys, often dry, cut in the river-lain and often pebbly red Triassic Bromsgrove Sandstone (around SO 735 307). (The Bromsgrove Sandstone has recently been renamed as the Helsby Sandstone).[14] This suggests that this rock, with the underlying Bridgnorth Sandstone and perhaps subsequent strata too, may have been deposited over much of Herefordshire but has subsequently been stripped from areas, such as Herefordshire, which either did not subside further to accumulate thicker sequences (as did the Worcester Graben) and/or were later uplifted.

Helsby Sandstone has been used as a building stone in the county. An example is in the church at Whitbourne (SO 725 569).

The border of Herefordshire takes in a very small area of Upper Triassic Mercia Mudstone near Great Malvern (around SO 762 487), part of the great thickness deposited in the Worcester Graben and which today floors the western half of the Severn Vale. This is a fine-grained red silty mudstone, with green-grey layers and spots, laid down possibly on a broad flat coastal plain or coastal lagoon/lake system. Evidence of the minerals gypsum (calcium sulphate with water of crystallisation) and rock salt (sodium chloride) is often found, together with desiccation (mud-) cracks, suggesting hot and dry evaporating conditions. There is no increase in the grain size of this deposit as the Malvern Fault is approached. This suggests that there existed no late-Triassic equivalent of the Malvern Hills, with attendant screes and flash flood deposits nearby to provide a coarser component (as seen for instance in the Bristol region), but that the whole area, including the Hills (presumably buried at this time) may have been part of this featureless plain. Once again, the implication is that the former, probably thinner, cover has been eroded away.

The same logic has been applied to the rest of the Mesozoic. In particular, the Lower Jurassic marine clays and limestones are thought by some to have been deposited right across Wales, as was probably the Upper Cretaceous Chalk. Unfortunately, little or no direct evidence has yet been found to support or detail these ideas. Some authorities believe that the various level surfaces suggested by the accordance of summit heights, such as the surface on which the Wye meanders first developed, represent the surfaces of erosion cut immediately prior to the deposition of such sediments and now exhumed by erosion of these relatively weak rocks. Nevertheless, it is certain that during the Jurassic and Cretaceous Periods, from about 200Ma to 66Ma, southern England was for most of the time submerged. Only with the start of the Cainozoic Era did Herefordshire once again become land and begin to evolve into the county we know today.

Further reading
Dreghorn, William *Geology Explained in the Forest of Dean and Wye Valley*, (1968). The 2005 reprint by Fineleaf Press is mainly concerned with the area to the south, but this is a very well written account. The general chapters are applicable to our area and there are three

specific chapters – on the Symonds Yat, Ross to Kerne Bridge and Hope Mansell-Wigpool areas. It was written before knowledge of plate tectonics so some of the interpretations are somewhat dated but it is still the best introductory volume on the area.

The H&W EHT's trail guides for *Wye Gorge* and *Symonds Yat* cover parts of the northern Forest of Dean. BGS Map sheets 215 (*Ross-on-Wye*), 216 (*Tewkesbury*) and 233 (*Monmouth*) and the associated written publications cover the area in question. (See pages *xix-xx*, 'EHT and BGS Publications'.)

9 THE CAINOZOIC ERA – BEFORE AND AFTER THE ICE

The present topography of Herefordshire

Previous chapters of this book have given an account of the nature and the evolution of the rocks which form the underlying geology of the county. Some rocks have been described as being more resistant to erosion than others and so may be expected to form the hills in the landscape. However, as will be described in this chapter, it is only in the Cainozoic Era that the county's land form has reached its present state. The present surface rocks were buried under a great thickness of Cretaceous, Jurassic and, probably, Triassic sediments as described at the end of the previous chapter. In the early Cainozoic these sediments were removed by erosion, yielding an area of low relief. From this 'flat' land the present hills and valleys have been derived by further erosion. This chapter therefore discusses the processes by which the present form has been achieved, and starts by summarising the county's topography today.

The topography of Herefordshire (Figs. 1.2 and 1.3) is shaped by the underlying diversity of its rocks and the early developments on the Cainozoic surface. As described in Chapter 1 and throughout the book, the county has a central plain developed on the lowest rocks of the Old Red Sandstone, the Moor Cliffs Formation (previously called the Raglan Mudstone, uppermost Silurian in age). Areas of higher land in the plain, for example Credenhill, Garnons Hill and the Bromyard Plateau, are capped by the lowest Devonian rocks, the Freshwater West Formation (previously called the St Maughans Formation). Surrounding the central plain are areas of rocks which are more resistant to erosion and so stand as higher land. Along the eastern border of the county are the Precambrian Malvern Hills and Silurian limestones pushed up along the Malvern Axis, in the Woolhope Dome area and on Shucknall Hill. In the north and west, there are Silurian hills along the line of the Church Stretton fault system and other major fault lines. The southernmost part of Herefordshire has the Carboniferous limestones and the Upper Devonian Huntsham Hill Conglomerate (previously called the Quartz Conglomerate) of the Wye Gorge and the Forest of Dean Plateau. In the south-west are the Black Mountains underlain by the more resistant rocks of the highest part of the Lower Devonian, the Senni Formation and the Brownstones Formation, mostly sandstones. Thus, the oldest exposed rocks are generally in the north of the county and the youngest are in the south, in line with the overall tilt of southern England to the south-east, as described later in this chapter. (The Malvern Hills present a clear exception to this, of course.) The main rivers are the Wye and its tributary the Monnow, the Lugg, with its tributaries the Arrow and the Frome, and the River Teme.

Erosion mechanisms and their influence in Herefordshire

The various rocks are described as being more or less resistant to erosion. The most resistant rocks are the igneous rocks with their interlocking crystals, or sedimentary rocks in which the grains are well cemented together and are composed of strong materials such as quartz. The forces of erosion are several in number, the principal ones being chemical solution in water (particularly applicable to limestones and lime-cemented rocks such as many types of sandstone) and mechanical breakage followed by movement under gravity. Mechanical breakage may be due to earlier tectonic movements, such as those along the many faults in the county. These lead to lines of relative weakness which are more susceptible to erosion. Mechanical weakness is more generally caused by large diurnal temperature fluctuations either in a hot desert or in periglacial conditions, when freeze/thaw action (water freezing and expanding in pre-existing cracks and later thawing) is especially potent. Abrasion by stones carried as the load in fast-flowing rivers or at the base of glaciers is effective in deepening valleys.

The more resistant rocks, most sandstones, siltstones and igneous rocks, generally make the highest hills. They form the high hills which lie on the boundary of Herefordshire on all sides, enclosing the low central plain. The greatest altitude in Herefordshire is 703m OD, on the Welsh border about 2km south-east of Hay Bluff, close to the north-eastern escarpment of the Black Mountains, formed of sandstone. The sand grains are of silica, which is chemically inert, while much of the cement is iron oxide with some calcite. Only the latter is easily dissolved in water and the rock is not highly porous. These properties make the rock resistant to erosion.[1] The next highest peaks in the county are in the Malvern Hills (425m OD). Here, the igneous rocks consist of interlocking crystals of insoluble silicate minerals which are very strong and show rather small thermal expansion, which renders them resistant to temperature fluctuations. However, these rocks have been much cracked by ancient tectonic forces and the cracks have rendered the rock susceptible to further breakage by freeze/thaw action. The rock fragments produced have been moved away from the hills by solifluction, the slow downward flow of water-saturated material under periglacial conditions. Even this may not have a major effect; a rough calculation by Phillips suggests that the height of the Malvern Hills was reduced by only about four metres in the most recent phase of the Ice Age.[2]

The Silurian hills of north Herefordshire attain altitudes of about 370m OD in Mortimer Forest and even greater in the west, on Hergest Ridge and Bradnor Hill. These rocks contain bands of limestone but are chiefly composed of mechanically tough siltstone, usually in a massive form and so are largely impermeable. Thus, although containing carbonate cement, these rocks are rated as erosion-resistant. Similar comments apply to the lower ridges of Silurian rocks near Malvern and in the Woolhope area and to the Carboniferous rocks of the northern Forest of Dean (about 220m OD). The Carboniferous limestones in the south of the county are susceptible to chemical solution in water penetrating joint planes and this leads to the development of the vertical cliff faces seen in the Wye Gorge. By comparison, the sandstones elsewhere are more susceptible to weathering processes which reduce the gradient of the valley sides and produce valleys that are more open.

The hills within the central plain rise to about 240m OD and consist of sandstones (with some mudstones) of the Freshwater West Formation. These rocks contain appreciable amounts of carbonate as cement. The sandstones are generally permeable allowing dissolution of the cement but the main mineral present, the silica of the sandstone grains, is insoluble so the rock maintains some strength. A calcrete band, the Chapel Point Limestone, underlies the steep slope where the land rises above the mudstone (of the Moor Cliffs Formation) which forms the low ground within the Herefordshire Plain. This is occupied by the flood plains of the major rivers. The mudstones become fragile when wet and this can spur erosion in the form of landslips.

The Palaeogene and Neogene Periods
The dawn of the Cainozoic Era, 65 million years ago, saw the elevation of most of what is now Britain into a land area. Uplift was greatest in the north and west.[3] This is attributed by many to the splitting apart of Europe and Greenland but some areas, notably southeast England, the Western Approaches and the North Sea, continued to sink and receive sediments. At first these sediments tended to be quite coarse-grained, reflecting erosion of the uplifted landmass, but from about 50 million years ago (approximately the end of Palaeogene volcanic activity in the north-west) they tended to be fine-grained muds and limestones, indicating lack of erosion and probably a subdued land surface with slow-flowing rivers (Fig. 9.2).[4] Was this the period that saw the Wye create its impressively large meanders (which developed on a level plain near sea level when rivers flowed through their own sediment and cut sideways rather than down)?

It is not known how much of Wales and the Marches was formerly covered by Mesozoic sediments but these are now found only outside the area of Palaeozoic rocks in Wales and the Welsh Borderlands. Some authorities suggest that the remarkably level skylines at various levels in many parts of Herefordshire and mid-Wales show the stages of the erosion of the original uplifted surface.[5] In some places, drainage patterns take no account of the underlying rock types and geological structure. The rivers possibly developed their courses on weaker Mesozoic rocks, which have since been removed by erosion, and have maintained their early courses.

This situation persisted, with minor interruptions due to Alpine earth movements, centred far to the south of Europe but which also affected Britain by varying amounts. The rivers tended to maintain their early meandering courses in many cases though now cutting down into the older, newly uplifted rocks. This is known as superimposed drainage and is shown especially well in the Wye Gorge area as detailed below. Differential uplift was very

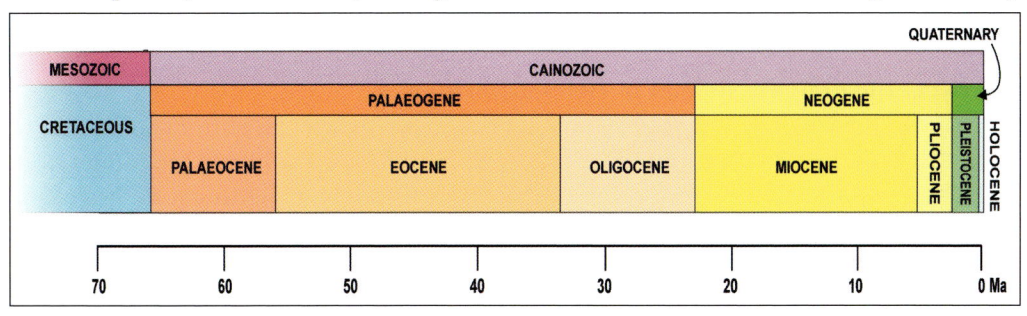

Figure 9.1 The Cainozoic time scale. In outdated nomenclature, the 'Tertiary' period comprised the Palaeogene and Neogene Periods. (Drawn by John Payne)

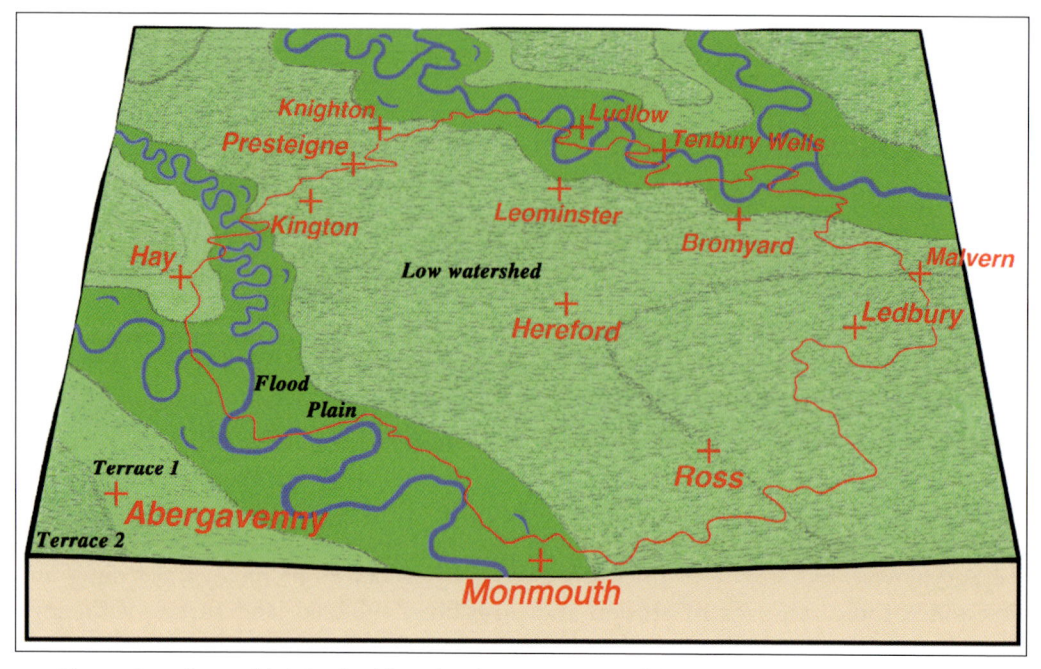

Figure 9.2 A possible Herefordshire landscape in most of the Cainozoic Era (from perhaps 60Ma). Large meandering rivers brought mud and sand south-east from the higher ground which lay to the north-west, across a low-lying, densely vegetated region (similar to South-East Asia today). The position and form of the rivers is conjectural but the source of the proto-Thames is believed to have been in north or central Wales and the river may have flowed across our region. (Drawn by Gerry Calderbank from a conceptual sketch by Dave Green)

marked in some areas where old faults allowed it, such as in the contrast between the elevations of Snowdonia and Anglesey. The development of the modern landscape can probably be attributed to differential erosion of such an upraised surface over the past 2.5 million years, that is, during the Quaternary Period.

The uplift imposed a dip generally towards the south-east and the rivers tended to flow in this direction. The sources of the Rivers Severn and Wye are on the eastern slopes of Plynlimon, whose summit is only 24km from the sea at Aberystwyth. These rivers follow a very long route to the east and south eventually reaching the sea in the Bristol Channel. There have been many alterations to the original drainage pattern.

The rivers and drainage systems in Herefordshire

Much of the drainage in Herefordshire follows the general regional trend from north-west to south-east (Fig. 1.2). In contrast to this, ancient long fault lines cut across the county from south-west to north-east and provide more easily eroded paths at a right angle to the general trend. Thus, near Hay-on-Wye, the River Wye follows the Swansea Valley Disturbance (Figs. 8.9 and 9.10). The Monnow south of Pontrilas follows the Neath Valley Disturbance and, further to the north-east, the River Frome just before its confluence with the River Lugg also follows this fault line. Other strong influences on the pattern of rivers, such as the effects of glaciation, are numerous and evident and are described later in the chapter.

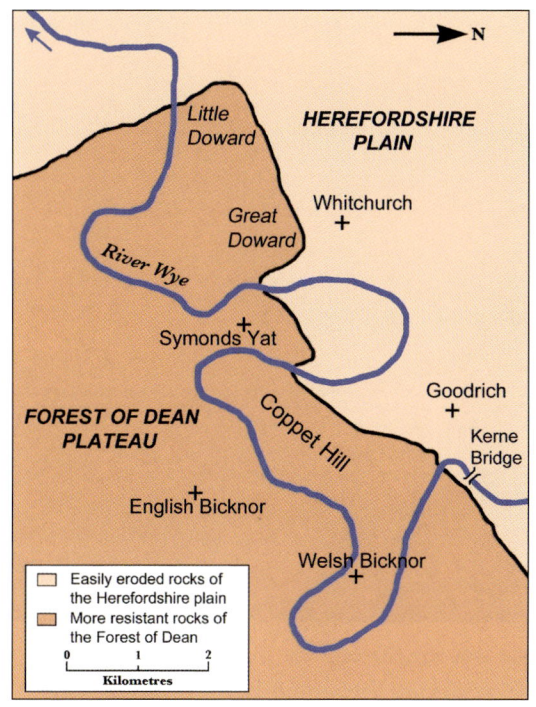

Figure 9.3 Map of the area between Goodrich and Little Doward. (Drawn by Gerry Calderbank from a conceptual sketch by Moira Jenkins)

The Wye Gorge; an example of superimposed drainage

Downstream from Kerne Bridge, south of Ross-on-Wye, the course of the River Wye bears no relation to the underlying geology. The river meanders several times, flowing alternately over the resistant rocks of the Forest of Dean Plateau, with spectacular examples of steep-sided gorges, and the weaker rocks of the Herefordshire plain, where there is a wider, shallower valley (Figs. 9.3 and 9.4). This is a well-known example of superimposed drainage.[6] If the river had developed its valley on the rock structure we see today, it would have taken the straightforward route and followed the line of the more easily eroded rocks of the Herefordshire plain to the north of the Forest of Dean.

The aerial photograph in Figure 9.4 shows the upper (western) two thirds of the area in Figure 9.3. You can see that the more resistant rock of the Forest of Dean Plateau stands up as higher land, which is

Figure 9.4 Aerial photograph of the Symonds Yat area looking south-west from near Goodrich with unwooded Coppet Hill in the foreground. (© Derek Foxton Archive)

Figure 9.5 The entrenched meanders of the River Wye are cut into the Carboniferous limestone plateau of the Forest of Dean. The view is northward into Herefordshire with, across the river, Great Doward to the right (east) and Little Doward to the left, separated by some fields in a dry valley. At the Biblins by the river is the youth camp site on the only area of flood plain in this section of the Wye Gorge. Symonds Yat is to the right, outside the frame.
(© Derek Foxton Archive)

mostly wooded, through which the river has cut a narrow deep gorge (Figs. 9.4 and 9.5). On the right in Fig. 9.4 is part of the Herefordshire plain, with the river flowing in a wide shallow valley around the meander by Huntsham Bridge. In the foreground is the unwooded part of Coppet Hill which is part of the Forest of Dean Plateau.

The Wye Gorge Story

Stage 1

It is believed that during the Palaeogene, the river meandered across a coastal plain. The river valley developed on a layer of gently folded underlying rocks and sediments, Mesozoic and Cainozoic in age, which have since all been removed by erosion. Older rocks at greater depth below the surface were more steeply dipping. At this time sea level was about 200 metres higher than at present (Fig. 9.6).[7]

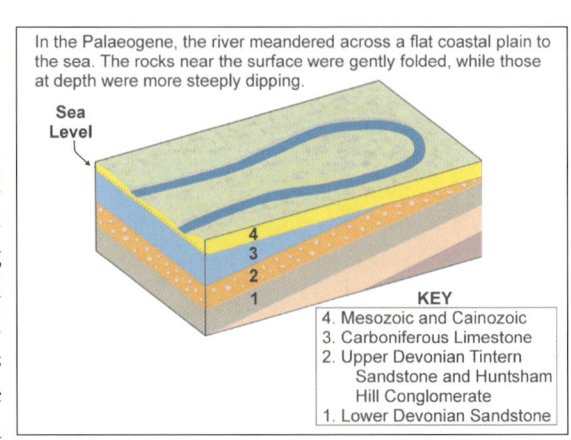

Figure 9.6 The Wye gorge – mid-Palaeogene
(© H&WEHT)

In the Palaeogene, the river meandered across a flat coastal plain to the sea. The rocks near the surface were gently folded, while those at depth were more steeply dipping.

KEY
4. Mesozoic and Cainozoic
3. Carboniferous Limestone
2. Upper Devonian Tintern Sandstone and Huntsham Hill Conglomerate
1. Lower Devonian Sandstone

Figure 9.7 The Wye gorge – late Neogene (© H&WEHT)

The uppermost rock layers were stripped away as sea level fell relative to the land.
The meander pattern was preserved as the river cut a deeper valley.

Sea level falling or Land rising

HEREFORD PLAIN
There is a wide, shallow valley where the meandering river flows over less resistant Lower Devonian sandstones.

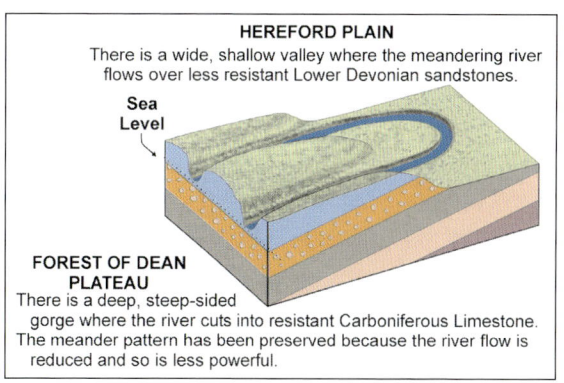

Sea Level

FOREST OF DEAN PLATEAU
There is a deep, steep-sided gorge where the river cuts into resistant Carboniferous Limestone. The meander pattern has been preserved because the river flow is reduced and so is less powerful.

Figure 9.8 The Wye gorge – present day (© H&WEHT)

Figure 9.9 Cave in cliffs near King Arthur's Cave (© H&WEHT)

Stage 2
During the late Cainozoic, as sea level fell (or the land was gently raised by earth movements), erosion of the land surface took place and the upper layers of rock and sediment were gradually stripped away revealing the more resistant rocks with different and more complex folds unrelated to the river's course (Fig. 9.7).

Stage 3
The Hereford plain, with a wide shallow valley and meandering river, has developed where the river flows over less resistant Lower Devonian rocks (Fig. 9.8). Relative changes in sea level since the Neogene and torrents of meltwater during the Ice Age caused the River Wye to cut downwards so fast that the original meanders were unable to migrate sideways and the river eroded a deep steep sided gorge in the resistant Upper Devonian and Carboniferous rocks of the Forest of Dean plateau, maintaining its meandering route. The present course is 'incised' into the Palaeozoic rocks.

Many other Herefordshire valleys are cut deeply into the original Cainozoic land surface. Much of this probably occurred during the Ice Age and so is described in that part of this chapter.

The King Arthur's Cave area
There is further evidence of a stage in the erosion of the land surface when the river was at a higher level than at present. Between Little Doward and Great Doward hills is a dry valley leading past cliffs in the Carboniferous Gully Oolite (formerly called the Crease Limestone). Along the cliff line is a series of water-worn caves, including

King Arthur's Cave (SO 5458 1558). The River Wye is now flowing at a level almost 100 metres below that of the base of these limestone cliffs, which have been undercut and smoothed by flowing water in the past as seen in Figure 9.9.[8] Extensive archaeological work has been carried out in these caves.[9]

Parallel drainage pattern in the south-west of the county

In the far south-west of the county, in the approaches to the Black Mountains and the Black Mountains themselves, is a second example of a drainage pattern established on the ancient uplifted land surface. Here, a series of rivers flow from north-west to south-east giving a good example of parallel drainage (Fig. 9.10). The drainage developed on the uplifted Neogene surface dipping from north-west towards the south-east and the streams flowed in the direction of greatest slope (consequent streams). The summits of the hills on the northern and eastern edge of the Black Mountains form a remarkably level surface, the remains of a much-eroded plateau. It is underlain by hard rocks of the Devonian Senni Formation and Brownstones Formation.[10] There is another lower surface on the less resistant sandstones of the Freshwater West Formation (St Maughans Formation) between the Black Mountains and the River Wye to the north-east.

The series of parallel valleys crossing this area includes the Dulas Brook and the River Dore in the Golden Valley. There are also the Escley Brook, River Monnow and the Olchon Brook. Across the border in Wales are the Afon Honddu, Grwyne Fawr, Grwyne Fechan and Rhiangoll. All the Herefordshire streams eventually drain into the Monnow which changes course at a sharp angle to flow along the major fault line of the Neath Disturbance and finally into the River Wye (Fig. 9.10).

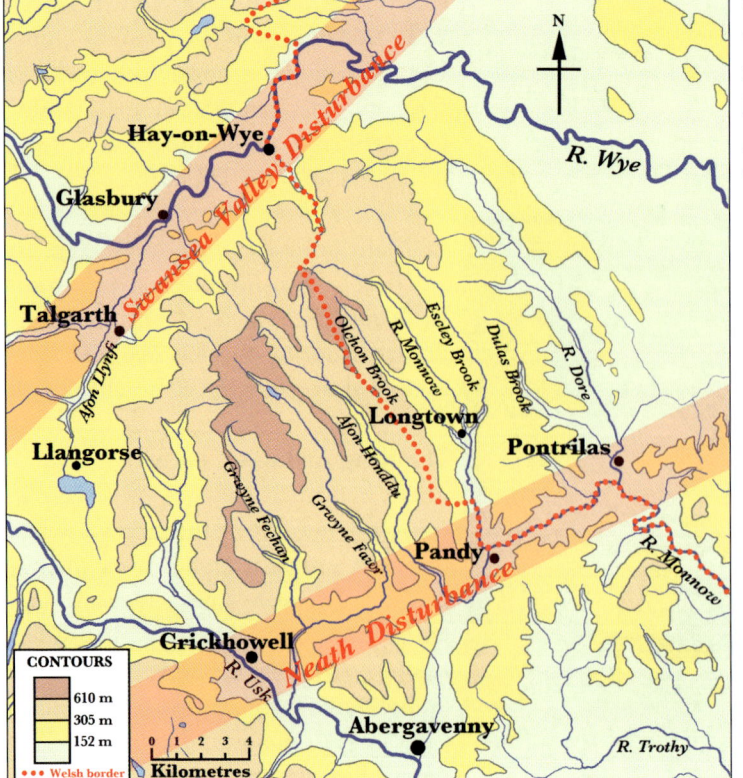

Figure 9.10 Map of an area of parallel drainage in the Black Mountains and south-west Herefordshire. (Drawn by Gerry Calderbank)

The Quaternary: Ice Age and Rivers in Herefordshire

The Quaternary Period

There is a great debate at the moment about global warming; whether it is taking place and to what extent human activity is affecting changes in climate. During geological time there have been many great changes in the Earth's climate, most recently in the Quaternary Period, during which ice sheets and glaciers advanced and retreated across the county and changed the routes of many rivers. The results of these events can be seen particularly in northern Herefordshire.

The Quaternary Period represents the last 2.6 million years and is subdivided into two Epochs, the Pleistocene and the Holocene, of which the Holocene occupies only the last 12,000 years (Fig. 9.1). Glacial episodes dominated the Pleistocene, each one generally lasting between 40,000 and 100,000 years. The very cold phases, during which glaciers and ice-sheets expanded across Britain, are known as glaciations. In interglacial periods, sea levels rose as the land-based ice melted, sea water expanded thermally and the country periodically enjoyed temperatures more typical of modern southern Europe. In these warmer phases, vegetation typical of cold climates receded towards the poles following the ice sheets, to be replaced by forest. While the short time-span of the Quaternary allowed few major new animals to evolve, climatic fluctuations meant that species had to migrate, evolve, or become extinct. It was under these testing conditions that the modern fauna and flora developed and, most notably, humans came to dominate.

Detailed information about temperature changes during the Pleistocene can be obtained by comparing the ratio between isotopes Oxygen 18 and Oxygen 16 in deep-sea cores. Water (H_2O), composed of O^{16} rather than O^{18} is preferentially evaporated from the oceans and some of it falls as snow on continents. At times when the build-up of snow and glacial ice occurs, there is therefore a corresponding enrichment of the oceans in O^{18} and a higher ratio of O^{18} in shells and hard parts of marine organisms. On this basis, the Quaternary has been subdivided into a series of Marine Oxygen Isotope stages to which each Glacial, Stadial, Glaciation, Interstadial and Interglacial stage can be correlated. (Stadial stages are periods of extremely cold climate usually associated with glacial advance. Interstadial stages have minor retreats of continental ice sheets during a phase of glaciation. They are shorter than interglacial periods, lasting between 50 and 500 years.)

During the later glacial stages in Herefordshire, advances of the ice-sheets have obliterated evidence of earlier glaciations. In Herefordshire today we can see deposits from two glacial periods, which used to be called the Older and Younger Drift. These are now classed as the Anglian (Middle Pleistocene) and Devensian (Late Pleistocene) Stages, about 450,000 and 20,000 years ago respectively. The deposits of this age have had insufficient time to be consolidated as rocks and so remain as 'superficial deposits' in the form of loose gravels and sands. It is by the study of these deposits and the present day geomorphology that clues about the landscapes of the past may be gained.

The Holocene Epoch began almost 12,000 years ago, after the last ice receded, and continues to the present day. It can be considered as an interglacial period of similar duration and temperature conditions to the many others that occurred throughout the preceding Pleistocene.

The Early to Middle Pleistocene landscape

Very little is known about Herefordshire's landscape in the early part of the Quaternary, much before about 450,000 years ago. Its detailed remains have been removed by the advance of the Anglian ice sheet. We can be much more certain about the early Middle Pleistocene, the Anglian Stage. Just before the advance of the Anglian ice, there were rivers that flowed north to south across the county (Fig. 9.11). These deposited the two oldest groups of Quaternary sediments in Herefordshire, in the Cradley Brook Valley and around the Bromyard Plateau.[11]

In the valley of the Cradley Brook, on the western margins of the Malvern Hills, there is evidence of a large, braided river during the earliest parts of the Middle Pleistocene cold stage.[12] The river, known as the Mathon River, deposited sands and gravels up to five metres thick. These were formerly exposed in three large gravel pits near Mathon (SO 735 458), now flooded or back-filled. The pebbles in the gravels are mostly of Silurian rock but also include igneous and metamorphic rocks derived from the Church Stretton area. This suggests that the river's catchment extended far to the north. The river itself is likely to have followed the western side of the Malvern Axis, including part of what now comprises the middle Teme Valley, north of Knightwick.

Simultaneously with the Mathon River, before the Anglian ice advance, a braided river system deposited gravels on the flanks of the high ground of the Bromyard Plateau, between 140m and 220m OD. These gravels contain small proportions of Triassic, Carboniferous, Precambrian and Lower Palaeozoic pebbles which were derived from the north. Gravels of the same age are found in tributary valleys of the River Lugg, such as the Humber.

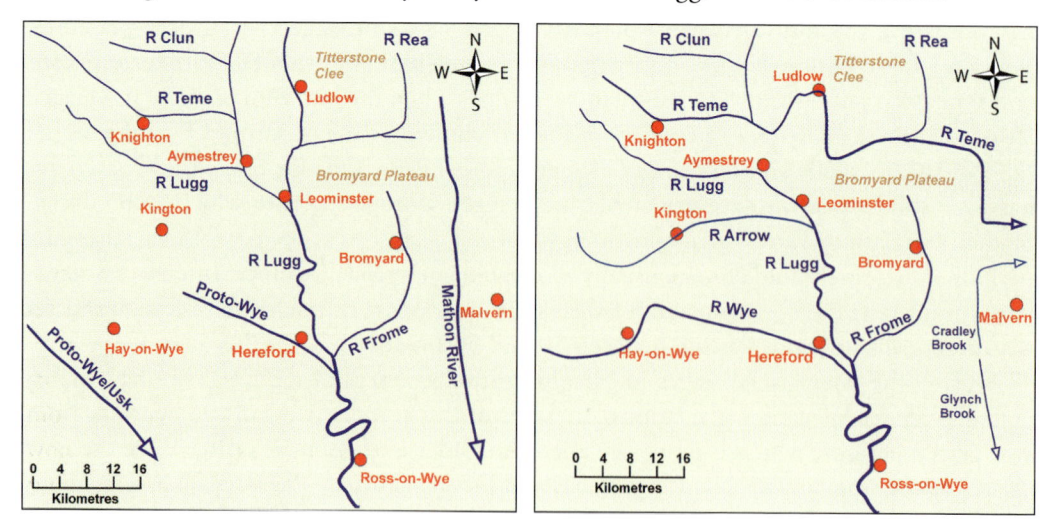

Figure 9.11 Middle Pleistocene drainage system in Herefordshire (left) compared with that of today (right). The effect of river diversions is very marked, especially in the north of the county although many segments of the river valleys remain the same.
(Left-hand diagram is modified from Richards 2007) (© A.E. Richards)

Incised meanders

The superimposed course of the River Wye in the Wye Gorge was described earlier. Many other Herefordshire valleys are cut deeply into the original late Neogene land surface by rivers which were rejuvenated in the Ice Age, given new vigour by sea level falling relative to the land and by the large discharge of water from melting ice sheets. The gravels found in the sections of valleys with incised meanders show that the topography was similar to the modern relief and many of the valley systems are of great antiquity. Such gravels are found on the flanks of the incised meander of the River Lugg near Bodenham (SO 534 512), where the Lugg has cut a deep valley through a line of hills making a huge loop around Dinmore Hill (see diagram on p.*viii*). The Sapey Brook (SO 70 58) and other streams have cut down deeply into the Bromyard plateau.

On a journey down the Wye from Hereford, different valley shapes are apparent, reflecting the variation in underlying rock type. Above Hereford and as far as Mordiford, the Wye flows over mudstone rocks with spreads of glacial deposits. Here the valley has a wide flood plain.

From the Capler view point (SO 591 325) the next stretch of the valley is seen. Here the meanders show classic interlocking spurs and an asymmetric valley profile as the river flows through the Devonian sandstones of the Brownstones Formation. These are called 'ingrown meanders'. Here, the sandstone rocks are susceptible to weathering, which reduces the slope of the valley sides. This is seen on the section of river from Capler to Kerne Bridge.

In contrast, where the rock is resistant limestone, weathering is far smaller. The rock is strong and there is little surface water to aid chemical weathering. (Rain water mostly runs

Figure 9.12 The Wye Gorge downstream of the Biblins (c.1890 to 1900).
This section of the gorge looks rather different today, being more overgrown.
(Library of Congress, Prints & Photographs Division, Photochrom Collection, ppmsc.08719)

away down joints within the limestone.) Because of the resistance to erosion of the limestone rocks, the river has eroded downwards more quickly than the meanders could migrate sideways. Thus, in the Symonds Yat and Biblins area of the Wye Gorge, the River Wye has cut down deeply into the Forest of Dean Plateau, forming a valley with vertical cliffs of Carboniferous limestone and little or no flood plain. (Figs. 9.5 and 9.12).[13] Here the river has retained its original meandering course. These are called 'entrenched meanders'.

These changing valley shapes are seen again as the Wye flows through Monmouthshire (Gwent) with gorges in limestone areas and less steep valley sides in sandstone areas. The sandstone sections of the Wye Valley are those in which meander cut-offs have developed, because here the meanders are able to migrate sideways (see Box 9.2).

Middle Pleistocene (Anglian) glaciation

In the Anglian glaciation from 478,000 to 424,000 years ago, an ice-sheet from the Welsh mountains advanced across Herefordshire as far as the Cradley Brook Valley on the west side of the Malvern Hills. At the same time, ice from the north filled what is now the Severn Valley to the east of the Malvern Hills. The remnants of glacial deposits which overlie the river gravels of the Humber and Cradley Brook Valleys provide evidence for the ice advance.[14] To the west of the Malvern Hills, the ice-sheet dammed the Mathon River, causing the formation of a large lake[15] and may have been responsible for major drainage diversion in the area. The Mathon River, flowing from north to south, no longer exists. In the Mathon area, the Cradley Brook now flows in the opposite direction, from south to north. It crosses the high ground of the Malvern Axis, north of the Malvern Hills, through a deep valley occupied now by the Knapp and Papermill Nature Reserve (SO 749 517). This valley pre-existed, and may have been cut by a former tributary of the Mathon River or perhaps an overflow from the Severn Valley glacier. In the Cradley Brook Valley, river gravels above the glacial tills and glacial lake clays contain no pebbles derived from the north; they are from the modern streams.

At the same time, an ice sheet covered the hills in the north-west of Herefordshire. Examples of erratics composed of a variety of rock types are found on the tops of hills such as Bradnor Hill (SO 283 585). These erratics are rocks which were transported in the ice, sometimes for hundreds of miles, and were dropped when the ice melted. The Whetstone (SO 2598 5675) is an example of a large erratic, composed of a distinctive gabbro, carried to the top of Hergest Ridge from Hanter Hill, in this case only a kilometre to the north-west (Fig. 4.19).[16]

In eastern Herefordshire, only remnants of the deposits left by the Anglian glaciation have survived, usually on ridges where they have not been eroded by the more recent Late Pleistocene (Devensian) glaciation or by rivers. These older glacial deposits show evidence of the conditions of the time, with delta fans and large boulders carried by meltwater torrents. For example, a phase of glacial re-advance dammed meltwaters on the westernmost margins of the Bromyard Plateau, near Stoke Prior (SO 520 565). This formed a lake that occupied the small river valleys of the precursors to the Stretford, Humber and Holly Brooks. Eventually, the water reached a critical level and rushed back underneath the ice-sheet, eroding the deeply incised gorge at Risbury Bridge (Hill Hole Dingle), cutting down six metres through the coarse gravels and a further two metres into the underlying rock of the Moor Cliffs Formation. Below Risbury Bridge (SO 540 550), the modern Humber Brook now gently trickles between large boulders in the base of the impressive gorge cut by the meltwater torrent.[17]

Middle to Late Pleistocene river network development

The Anglian glaciation appears to have been responsible for a number of changes to earlier Quaternary river patterns, for instance the changes to the Mathon River described earlier.[18]

Other small-scale drainage modifications occurred. Pre-existing valleys at Blackwardine and Eaton Hill, respectively south-east and east of Leominster, became infilled with glacial sediments, causing the Humber Brook to erode a valley south of its previous course. The Stretford Brook was deflected northwards, following the route that it occupies today.[19] These tributaries to the River Lugg were subject to further modification in the Late Pleistocene (Devensian) glaciation.

At the end of the Anglian glaciation, meltwaters carried gravels in braided river channels down the valleys. At the same time, as sea levels rose, gravels were deposited in the valleys which had been deepened during the glaciation. The valleys of the River Lugg, Wye and Teme contain flights of river terraces which represent at least four phases of sedimentation under cold climate conditions (Fig. 9.13). Only remnants of these terraces now remain. The highest terrace, the flood plain of the early river, is the oldest and the meandering river cut into this to form a lower and younger terrace. This process was repeated, cutting down and producing successively younger, lower terraces until it produced the valley of the present day. (See Box 9.1.) Gravels were often deposited on the terraces by floods.

The catchments of early rivers can often be found by studying the sources of the pebbles in the gravels they deposit. In the gravels on older terraces, pebbles from the Shropshire area show that the River Lugg's catchment area then included areas north and north-east of Ludlow. In later terrace deposits, these northern erratics decrease; there are many more locally derived pebbles as well as igneous clasts that are derived from Hanter Hill and Stanner Rock, just over the Welsh border north-west of Kington.

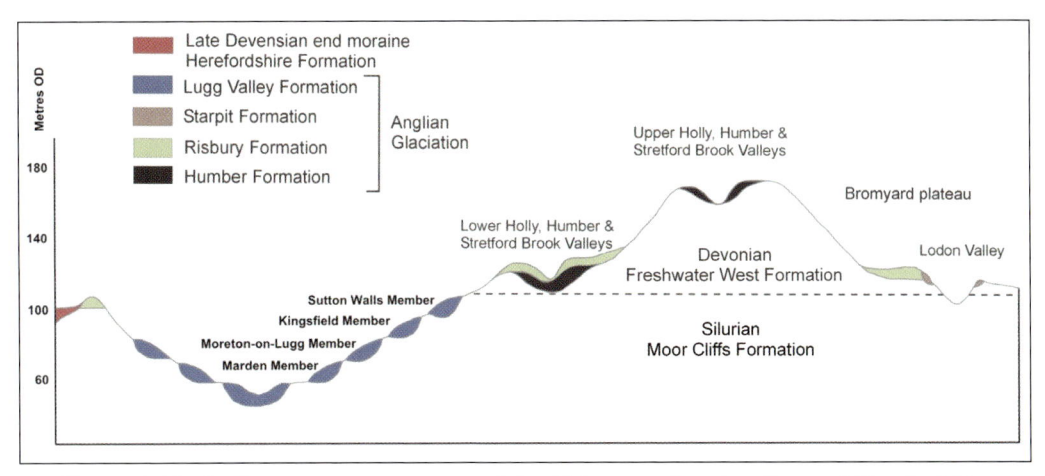

Figure 9.13 A schematic representation of the Pleistocene deposits of central Herefordshire. (Adapted from Fig. 5 in Richards 2007) (© A.E. Richards)

> **Box 9.1** **River Terraces**
>
> A river terrace is a former flood plain of a river into which the river has subsequently cut a deeper channel with its own flood plain. This can occur as a response to a fall in the base level for the river, perhaps the level of the sea. This process may be repeated over several cycles of base level fall, resulting in a 'staircase' of terraces as shown in Figure 9.14. Clearly, terrace no. 3, the highest, is the oldest and terrace no. 1 is the youngest.
>
> *Figure 9.14 River terraces. (Drawn by John Payne)*

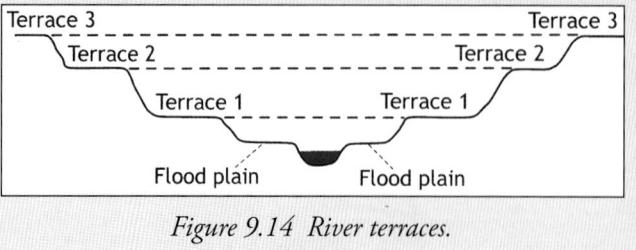

The Wye Valley also has four sets of river terraces, each with gravels deposited by a braided river in a cold climate. The oldest terraces, in the Wye Valley numbered 4 and 3 by researchers, occur at high levels in and around Hereford City and in the Wye Valley below Hereford. Terraces 1 and 2 occur throughout the Wye Valley below Hay-on-Wye. All the terraces are poorly exposed, with the majority of pebbles being far travelled, derived from Lower Palaeozoic outcrops in the Upper Wye basin of central Wales.

Fossils are remarkably rare in the gravels; the only finds being of the horse, *Equus fossilis*, and the rhinoceros, *Rhinoceros tichorhinus*, from former pits in terrace 4 in the Lugg Valley.[20] The River Teme has a similar set of river terraces. Its development has been complex. The upper reaches of the Teme with the Ledwyche Brook and River Rea were once tributaries of the River Lugg.[21]

Late Pleistocene (Devensian) glaciation

The most recent glaciation, the Devensian, occurred relatively recently, in the Late Pleistocene. It began about 115,000 years ago and ended about 12,000 years ago, with a maximum ice extent between 23,000 and 18,000 years ago.[22] The Devensian ice-sheet spread across the Hereford Basin, fed by ice fields covering the Welsh Massif and entering the area principally as a glacier flowing down the Wye valley. This glaciation finally shaped most of the topography and the river courses we see today.[23] In the north it pushed up against the hills of Silurian limestone and siltstone that run from Kington to Orleton[24] and in the south against the Devonian (Lower Old Red Sandstone) escarpment running from Hay-on-Wye to Whitfield (SO 42 33). The ice moved from west to east, reaching its maximum extent near the current Lugg Valley to the north of Hereford.

Figure 9.15 shows the ridges of moraine which mark the stages of the retreat of the Late Pleistocene ice. At its maximum extent, it reached as far east as the racecourse in Hereford, Burghill and Tillington, the flanks of Dinmore Hill and Stoke Prior. It also produced a large arc-shaped moraine at Orleton.

The most extensive current exposure in the Late Pleistocene end moraine is seen in a gravel pit at The Leasows (SO 537 579), on the flanks of the Stretford Brook (Fig. 9.16). Here, over six metres of heavily deformed glacial deposits are exposed.

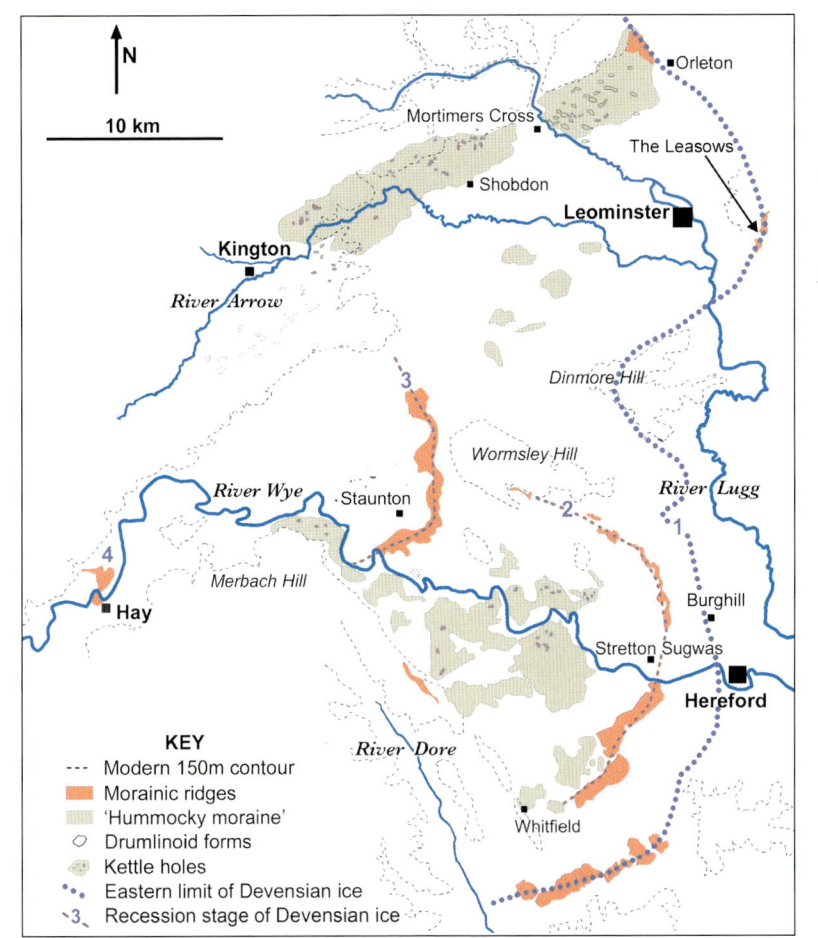

Figure 9.15 The moraines and associated deposits of the Late Devensian ice-sheet in Herefordshire. (Adapted from Fig. 9.6 in Richards 2005) (© A.E. Richards)

Figure 9.16 Late Devensian end moraine seen at the Leasows. (© H&WEHT)

At the end of the Devensian, the ice began to melt and the glacier retreated towards the Welsh massif. In the Wye valley there were several stages of still-stand when the rate at which the glacier advanced down the valley matched the rate at which the ice was melting so that the front of the glacier remained at the same place for some time. At each location where the ice front halted, eroded debris carried by the glacier accumulated to form a ridge of moraine. These can be seen at Stretton Sugwas, Staunton on Wye and between Hay-on-Wye and Clyro (Figs. 9.15 and 9.16).

Immediately behind (west of) the ridges there are extensive spreads of glacial deposits including hummocky moraine with small but remarkably deep hollows. As the ice front retreated, blocks of ice became buried in sediments formerly carried by the ice-sheet. When these ice blocks melted the unconsolidated material above collapsed forming hollows, known as 'kettle holes', many of which now contain small lakes. One at Moccas contains over eight metres of lake sediments. Organic detritus from sediments in kettle holes at Bridge Sollers (SO 421 417) and Stansbach (SO 343 607) has been dated, giving radiocarbon dates of 15,411 to 13,841 and 15,196 to 13,499 years before the present respectively, the youngest possible date for melting of the ice in the Hereford Basin.[25] The fields near Kenchester contain several kettle hole pools in a small area (SO 435 425).

A lobe of the Wye glacier moved down the Arrow Valley pushing against the line of hills from Bradnor Hill (SO 280 585) past Wapley and Shobdon Hills and the Croft Ambrey ridge (SO 444 668). As the ice melted, it left sediments banked against the hills from Kington to Orleton. Dr A.E. Richards has pointed out that this continuous stretch of moraine takes a variety of forms. In the Upper Arrow Valley there are hummocky features, some of which have rocky cores and often contain large angular boulders. Near Kington, the moraine is defined by an irregular, hummocky topography commonly with kettle holes, such as those near Titley and Flintsham. Near Shobdon, the hummocks look like drumlins, the 'basket of eggs' topography of sediment moulded by movement of the ice and meltwater moving under the ice (Fig. 9.17). East of Mortimer's Cross, near Lucton, there are more drumlinoid forms that are aligned roughly from south-west to north-east. To the south, at Stretford, a series of ridges marks the position of a series of eskers. Eskers are ridges of gravels deposited by meltwater in channels under the ice as it melted.

In the north-west of the county there are clear examples of deep steep-sided channels cut by meltwater, many of which now have no stream. Examples of these are seen on the south-east slopes of Hergest Ridge (SO 263 563; Fig. 9.18) and Bradnor Hill (SO 289 579) and in the Fishpool valley (SO 451 660) near Croft Castle. Just to the north of the Arrow

Figure 9.17 View of drumlinoid forms west of Shobdon looking south. The 'basket of eggs' topography is glacial moraine shaped under the ice. (© H&WEHT)

Figure 9.18 The steep-sided meltwater channel on the south-east side of Hergest Ridge is shown in this view looking west-north-west. The channel is marked by trees in its lower part but high on the Ridge it is on the bare hillside. The water drained into the River Arrow valley (in the foreground) near Hergest Court. (© Derek Foxton Archive)

valley near Huntington, is a distinctive arc-shaped valley with steep sides. This is known as Hell Wood Channel or the 'Rainbow' (SO 258 532) (Fig. 9.19).[26] Because of its slightly humped profile half way along its length, this is thought to be a sub-glacial channel eroded by debris-laden meltwater under pressure at the margin of the glacier when the valley, just to the south, was choked with sediment which is seen today as hummocky glacial moraine.

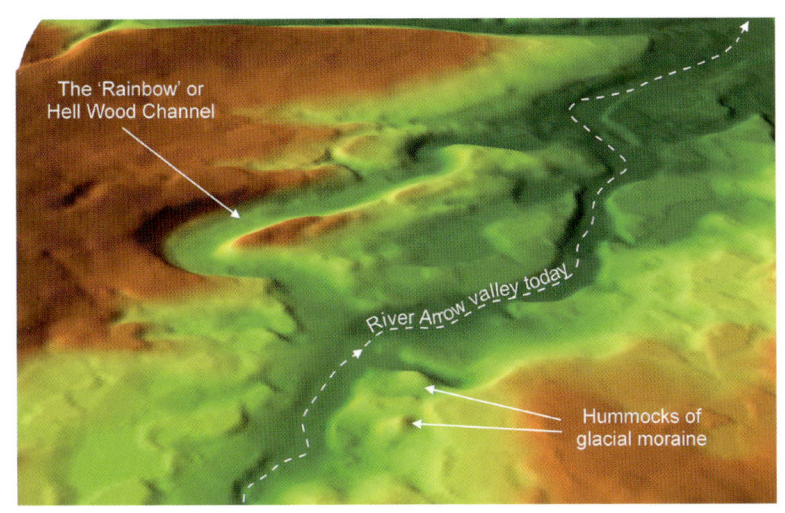

Figure 9.19 3D representation of the 'Rainbow' sub-glacial channel. (Contains Ordnance Survey open data © Crown copyright and database right 2013. Diagram produced by Dr M.J. Payne using NERC LIDAR data.)

Figure 9.20 (above) This photograph shows the view east-north-east from Wigmore Castle (SO 408 693) across the flooded Leinthall Moor to the Silurian limestone hills of Bringewood Chase. The flat ground of the Moor lies in the Wigmore Basin in the centre of the eroded Ludlow Anticline and is the floor of the ancient Wigmore glacial lake. (© John Payne)

Figure 9.21 (below) Aymestrey village is in the centre of this photograph, taken looking south-east. The River Lugg enters the picture from the bottom right, joins the Aymestrey valley and then flows south (diagonally to the right) towards Mortimer's Cross. The valley was once the southward course of the River Teme until it was blocked from the south by a lobe of the Wye glacier, causing the river's diversion to its present course through the Downton Gorge. The glacier left a ridge of moraine (bottom left corner of the frame). (© Derek Foxton Archive)

Blockage by glaciers and moraine caused changes in the courses of several rivers locally. The example of the Mathon River has already been mentioned. The course of the River Wye has also been altered. Its channel is believed to have formerly passed Staunton on Wye between Oaker's Hill and Garnons Hill and then to have followed a course close to that of the present-day Yazor Brook to Hereford. The Yazor Brook is now a small misfit stream (misfit streams are explained later in this chapter) in a broad valley floored by till and fluvioglacial gravels. Due to the blockage of its channel by the Stretton Sugwas moraine, the River Wye has cut a new route through a gorge-like valley at Breinton (SO 452 400).[27]

The course of the River Teme was blocked in the Devensian by a tongue of the Wye Glacier near Aymestrey (SO 425 655). A proglacial lake developed in the Wigmore Basin (Fig. 9.20). Meltwater from the glacier spilled northward through the gap in the hills at Aymestrey (Fig. 9.21). This built a delta at Yatton, extending northwards into glacial Lake Wigmore, whose level attained a height of 128 to 131m OD. At this height the lake water reached a low point in the topography and meltwater, loaded with gravels, overflowed cutting the present gorge of the Teme at Downton (SO 435 735) (Fig. 9.22).[28] (An alternative theory is that the Downton gorge developed subglacially).[29] The River Teme now flows to Ludlow and eventually joins the River Severn instead of following its previous course via Aymestrey to the River Lugg as shown in Figure 9.11.[30] The delta gravels have all been quarried from the Yatton gravel pit. A line of glacial moraine can still be seen as a steep-sided ridge (SO 428 658).

Similarly, a proglacial lake formed in the Presteigne Basin when the valley of the River Lugg, which previously had a route through the line of hills between Wapley and Shobdon

Figure 9.22 (below) The Downton Gorge follows the jagged line of trees across the centre of this photograph, which is taken pointing north-east. The cliffs in Aymestry Limestone at the south-west end of the gorge are seen amongst the trees in the foreground. It was here that glacial meltwater cut the channel though the Silurian rocks, diverting the River Teme into its present course through Ludlow.. (© Derek Foxton Archive)

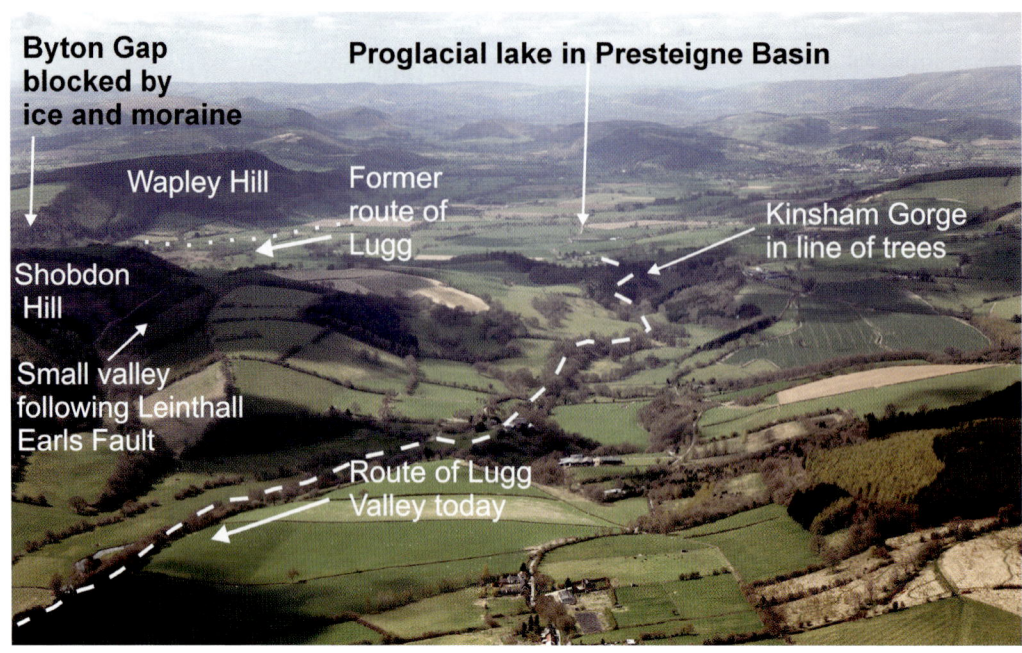

Figure 9.23 (above) Diverted route of the River Lugg, view looking west from Shobdon Hill. (© Derek Foxton Archive)

Figure 9.24 (below) The picture shows the River Lugg flowing from right to left (west to east) in the Sned Wood Valley, which is to the west and upstream of Aymestrey. The view is directly to the south and shows the Covenhope Valley, a valley now almost dry. This was the former route of the Lugg to Mortimer's Cross before it was blocked by ice. The Sned Wood Valley has been eroded in the rocks weakened by the action of the Leinthall Earls Fault along a nearly straight line through north Herefordshire. (© Derek Foxton Archive)

Hills, was blocked by ice and moraine (SO 37 63) near Byton. The River Lugg has a new course through Kinsham Gorge (SO 365 650) (Fig. 9.23) and Sned Wood Gorge (Fig. 9.24).[31]

Landslips and protalus ramparts

Landslips often occur where ground water seeps through permeable rocks overlying impermeable rocks or where rock layers are dipping steeply down slope. In areas subject to landslips, these have often been more frequent during the Pleistocene, when the superficial deposits were less stable in the periglacial conditions. Herefordshire provides many examples.

Around the edge of the Black Mountains on the south-western border of Herefordshire are many examples of landslips. The Ffynnon Limestone, a band of calcrete (a concentration of limestone in a fossil soil) outcrops as a line of crags on the steep slopes between the Senni Formation of the plateau surface above and the Freshwater West Formation (previously called the St Maughans Formation) of the lower land (Fig. 2.8).[32] Landslips are associated with the springs emerging at this horizon and probably took place during periglacial periods.

At Black Darren (SO 296 297) (Fig. 7.1), around the picnic area, the hummocky ground is the result of landslips. Above the landslip area is another feature, ridges running parallel to the valley side, formed under periglacial conditions and composed of large angular rocks.[33] Towards the end of the Ice Age, frost-shattered debris moved down the steep slopes over semi-permanent accumulations of snow, high on the valley sides. When this snow melted, the debris was left as a ridge, which is steep-sided on both the uphill and downhill sides. This is known as a protalus or pronival rampart (Fig. 9.25). More examples are seen over the Welsh border in the Honddu Valley.

Figure 9.25 (below) Protalus rampart at Black Darren. (© R. Bryant)

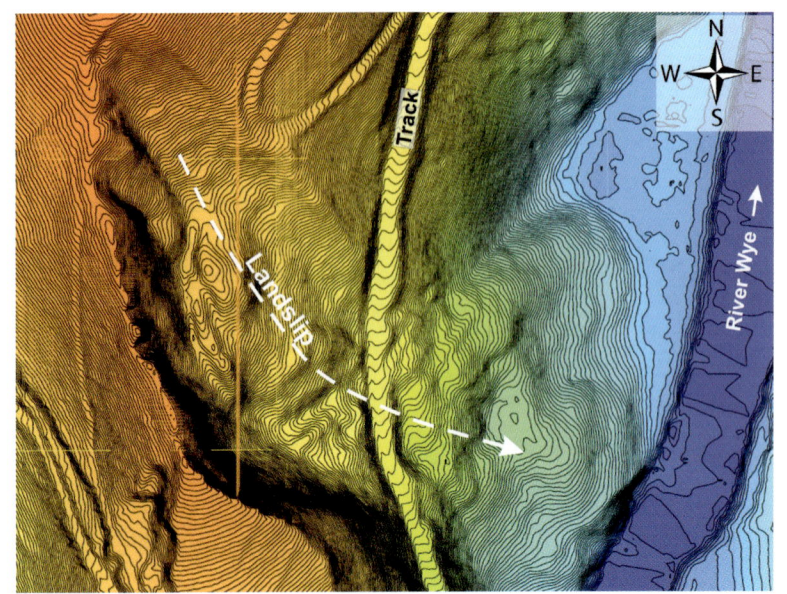

Figure 9.26 Landslip on the east side of Huntsham Hill shown by a contour plot at 50cm intervals. (Generated by John Payne from UK Govt. LIDAR data)

Several landslips have occurred in the Wye valley upstream of Symonds Yat. On the north and west sides of the river the rocks dip generally towards the centre of the Forest of Dean, that is, towards the river. Where the river has undercut the hillside, the higher strata may lose their support and slip towards the river. There are two such slips on the east side of Huntsham Hill. The southern one is depicted in Figure 9.26, where the lobes of slumped material are clearly seen (SO 564 164). The debris has reached the river and caused a slight diversion to the east, creating a small area of flat ground downstream. This small diversion is well seen from the famous viewpoint on Symonds Yat Rock (Fig. 2.24).

Although they were probably most frequent in the periglacial environment at the end of the last glaciation, landslips have occurred locally in historical times up to the present day. Around the outer edge of the Woolhope Dome, because of the geological structure, the bedding dips downhill parallel to the slope of the ground. Within the Silurian limestones, layers of volcanic ash have been chemically altered to form bentonite clay bands. These bands are impervious to water and become lubricated, providing a slippery, very weak layer of clay on which landslips have occurred. Examples of these are the Wonder (SO 633 365) (see also Chapter 10) and the Slip near Dormington (SO 592 400). The latter took place as a single abrupt movement on 15th March 1844 and was reported in the *Hereford Journal* of the 20th March 1844. The Slip is 290m long and up to 80m wide consisting of jumbled blocks of Aymestry Limestone up to five metres across.[34] Work ceased in the nearby Perton Quarry for a while after a similar landslip in 1979.

In the Dudale's Hope valley (SO 567 517) there is a landslip within the Freshwater West Formation. Groundwater penetrated the pervious layers of calcrete and sandstone and lubricated the underlying mudstones. This led to a slip in a belt 1.3 km long. Similar events took place in the winters of 1946-7 and 2013-4 near Shortwood Farm (SO 591 509) (Fig. 10.11) affecting 1.5 hectares and moving mature trees to a new position but still upright.[35]

River and drainage systems in Herefordshire today

The expansion of the Devensian ice-sheet into the Herefordshire Basin was the last major influence on the drainage of Herefordshire. The main effect of Devensian glaciation seems to have been further excavation of the Wye and its tributary systems, with the Wye becoming the dominant river system in the area because of drainage diversions in the northern parts of the county.

The landscape of Herefordshire is continuing to evolve. The fluvial processes recorded in the rocks of Herefordshire are still taking place today and we can see many places where rivers are still actively eroding the landscape. There are also examples of former river channels which have been abandoned. Some low-lying areas are subject to flooding and attempts have been made to lessen its effects. Some of these changes and the drainage patterns seen today are described below.

Meanders, misfit rivers and oxbow lakes

The River Teme and much of the previous upper catchment of the River Lugg now flow to the east to join the River Severn near Worcester. The River Lugg is now a misfit river, meaning that it is far smaller than the river which originally carved out the valley which it occupies south of Leominster. Before the Ice Age, the River Teme joined the Lugg and both of these rivers, swelled with meltwater, were able to erode a far wider valley than the present flow of the Lugg could achieve. The River Lugg and its tributaries now drain a smaller area, with its headwaters derived from the hills west of Presteigne. From here, the Lugg follows the line of the Leinthall Earls Fault before it flows through the gap at Aymestrey to meet its ancient middle course just north of Leominster (Figs. 9.21, 9.23 and 9.24). Figure 9.11 shows the changes in the drainage.

Once a river starts to meander, a process begins of continuous erosion of the outside of each bend and deposition on the inside of each bend.(Box 9.2; Figs. 9.27 and 9.28). As a result, the meanders gradually move downstream. The wavelength of the meanders is related to the volume of water in the stream. There was a great deal of erosion when rivers were swelled by meltwater at the end of glacial periods. Now that there is less water in the streams, smaller meanders develop on the side limbs of the larger meanders from the past.

Figure 9.27 The River Lugg meanders across the Presteigne Basin near Byton. (© H&WEHT)

Figure 9.28 The River Teme meanders over the flat floor of the old glacial Lake Wigmore. The photograph was taken pointing south-east, just south of Leintwardine. In the foreground are seen gravel banks in the stream and, in the field, the traces of now-abandoned earlier meanders of the river. (© Derek Foxton Archive)

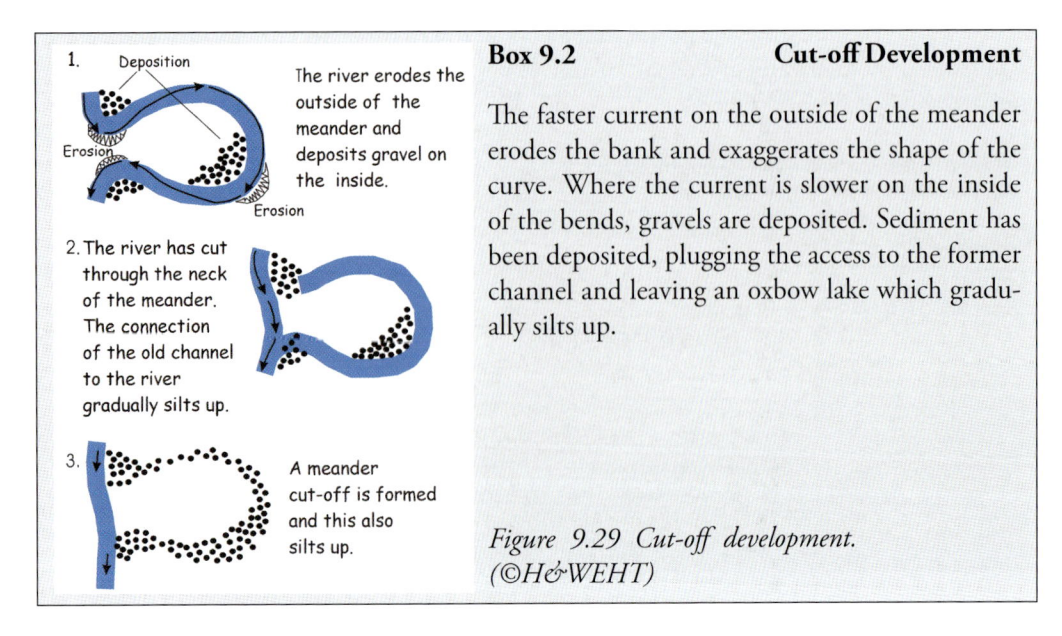

Box 9.2   **Cut-off Development**

The faster current on the outside of the meander erodes the bank and exaggerates the shape of the curve. Where the current is slower on the inside of the bends, gravels are deposited. Sediment has been deposited, plugging the access to the former channel and leaving an oxbow lake which gradually silts up.

1. The river erodes the outside of the meander and deposits gravel on the inside.
2. The river has cut through the neck of the meander. The connection of the old channel to the river gradually silts up.
3. A meander cut-off is formed and this also silts up.

Figure 9.29 Cut-off development. (©H&WEHT)

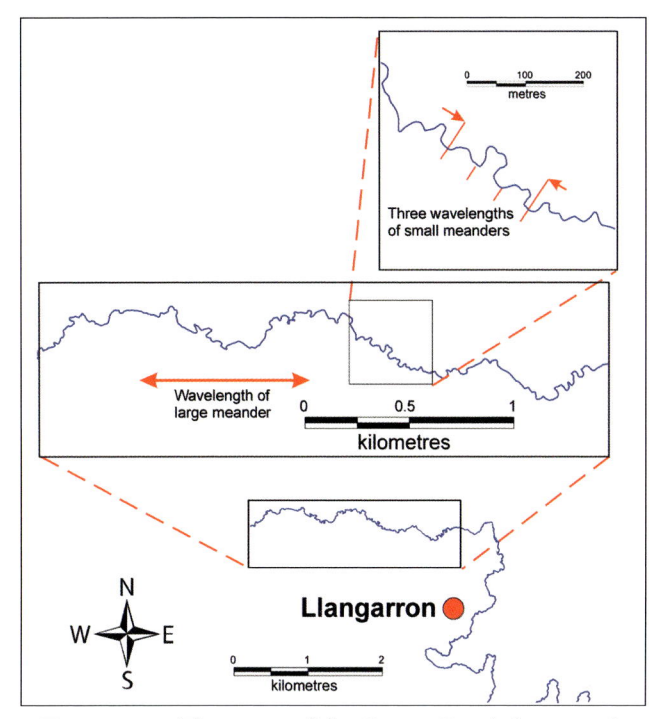

Figure 9.30 The course of the Garren Brook from north-west of Llangarron (SO 530 212) toward its confluence with the River Wye near Whitchurch. Meandering on two distinct scales is apparent. The two wavelengths are measured from the enlarged sections in the upper diagrams. They are 800m and 53m. (Drawn by John Payne)

This is well seen in the Garren Brook west of the River Wye near Ross-on-Wye (SO 53 22) (Fig. 9.30).[36] Here the wavelengths of both the old and the current meanders can be measured (800m and 53m respectively). The volume flow (the discharge) of a river when it is full to the top of the banks may be estimated solely from the wavelength of its meanders.[37] From this it can be shown that the old, large river carried about 200 times the flow of the present Garren Brook (when full to overflowing) and that its discharge was about the same as that of the River Wye today.

The modern rivers in Herefordshire show classic features of meandering streams. Fine examples of meanders and meander cut-offs, or 'oxbow lakes', appear along the River Lugg upstream of Byton in north-west Herefordshire (SO 36 63) (Box 9.2 and Fig. 9.27). Their formation is an active process which is continuing to change the river channel. The Lugg flows across the flat floor of the former proglacial lake in the Presteigne Basin. In this section of the river valley, there are examples of meander cut-offs where the river has cut through the neck of the meander, shortening its course. The parish boundary was drawn along the line the river used to follow. There is a telegraph pole, whose maintenance is the responsibility of Kinsham Parish Council, which is no longer on the Kinsham side of the river and no longer accessible without a journey to the nearest road bridge to cross the river. Other meanders in the Presteigne Basin have narrow necks, which the river will cut through in the not too distant future (Fig. 9.27).[38]

The River Wye enters Herefordshire just below Hay-on-Wye and for a few miles follows the line of the Swansea Valley Disturbance, a fault line trending south-west to north-east (Fig. 8.9). Below Clifford (SO 25 46), the river turns to flow eastward and meanders across a wide floodplain. The valley is underlain by easily eroded mudstones which are covered widely by spreads of glacial deposits or alluvium. Along this stretch of the Wye there are fine examples of meanders which are gradually migrating downstream as the river erodes the outside of the meander bend and deposits sediment on the inside of the bend. One such meander is seen at Lockster's Pool (SO 270 465) east of Clifford.

Abandoned meanders of the River Wye

The Wye valley shows evidence of historical meander cut-offs which have left long-abandoned meanders which are now well above the level of the present river. Figures 9.31, 9.32 and 9.33 show the area to the east of Ross-on-Wye, where there is an example of a former incised meander which is no longer the course of the River Wye. Chase Wood and Penyard Park (SO 61 22) are detached fragments of the Forest of Dean plateau. These hills have a similar height to the rest of the plateau and are capped by rocks of the Upper Devonian Huntsham Hill Conglomerate and Tintern Sandstone Formations. They are separated from the rest of the Forest of Dean plateau by a large valley, now only partly occupied by a small brook. This is the former course of the River Wye, which used to flow in a loop, east and south of Penyard Park and Chase Wood until it cut through the neck of its meander, shortening its course and leaving the wide valley seen today. The present day River Wye has continued to cut downwards as sea level has fallen relative to the land. The floor of the former valley is more than 30m above the height of the present day valley of the Wye.

Other examples of abandoned meanders are seen further downstream in Monmouthshire at Newland and Bigsweir.[39]

South of Bromyard in the Avenbury area, near the ruined church, is an example of an incised meander on the River Frome (SO 661 531). The former flat valley floor is seen on the surface of the interlocking spurs. The present-day river has cut down below this surface. Just south of this, at Hyde Farm, the former valley, now dry, can be seen just below the road, while the River Frome has cut a new straighter route just west of the farm (SO 666 522). This is an example of an abandoned meander on a smaller scale than that seen near Ross-on-Wye.

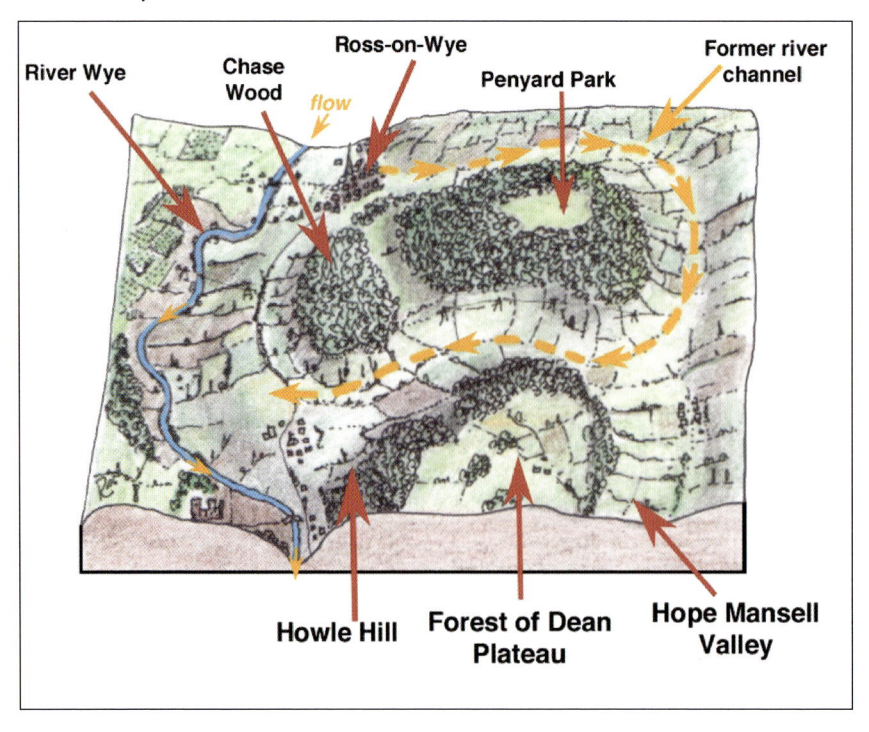

Figure 9.31 Diagram of an abandoned meander near Ross-on-Wye. (Drawn by Rollo Gillespie © H&WEHT)

Figure 9.32 (above) The picture illustrates the long-abandoned meander of the River Wye behind Ross-on-Wye (the area covered by Fig. 9.31 viewed from the lower right side). The view is to the north-west over Penyard Park and Chase Wood (in the middle distance, to right and left respectively) to Ross-on-Wye beyond. The abandoned meander of the River Wye (the dashed line) left the present course of the river (the dotted line) at Ross, ran on the north, then east (right) side of Penyard Park and then in the valley seen in front of it to rejoin the present river course south of Ross.(© Derek Foxton Archive)

Figure 9.33 (below) The southern end of Ross appears on the left of the frame. Chase Wood is in the centre and Penyard Park lies behind it to the left. Howle Hill is on the right. An old meander of the River Wye ran from Ross, behind Penyard Park and back to join the present course of the Wye south of Ross, emerging from the valley seen stretching eastward between Chase Wood and Howle Hill. (© Derek Foxton Archive)

Woolhope Dome, river capture and concentric rings of hills

In the Woolhope area (SO 60 37) Silurian rocks have been pushed up into a dome, elongated from north to south. Originally, a radial drainage pattern developed with streams flowing away from the higher land at the centre of the dome. However, most of the rocks of the Woolhope Dome are alternating limestones and shales. Tributary streams have found it easier to erode the weaker shales into valleys, leaving the more resistant limestones standing up as a series of concentric ridges (Fig. 6.13).[40] The Pentaloe Brook and its tributaries drain the northern part of the dome, finally emerging at Mordiford.

River capture

River capture occurs when one river is more actively eroding its valley than another nearby. When the watershed separating the two valleys is breached, the headwaters of the lesser stream flow into the river which has cut down more deeply.

There is an example of river capture in the Woolhope Dome (Fig. 9.34). To the east of Fownhope the Tan Brook has captured the headwaters of a stream which used to flow through the gap at Common Hill (SO 590 347), leaving a dry valley downstream.[41]

Another example of river capture is seen along the River Frome, a tributary of the River Lugg, which flows entirely within the county of Herefordshire. The Frome rises on the more resistant Freshwater West Formation rocks of the Bromyard Plateau and has cut a deep narrow valley in its upper reaches. Below Bishop's Frome (SO 664 484), the river leaves the plateau and the valley widens out on the weaker, easily eroded rocks of the Moor Cliffs Formation. Here the river is very muddy and has been described as the most turbid river in the country. This stretch of the river follows the line of the Neath Valley Disturbance between the upfolded Silurian rocks of Shucknall Hill and the Woolhope Dome.[42] Since the end of the Pleistocene, the River Frome has captured the River Lodon, which used to flow westwards to the north of Shucknall Hill to join the River Lugg. This former course of the Lodon is now a dry valley (SO 60 45) west of Monkhide, which was used as the route for the construction of the Hereford to Gloucester Canal. The old Lodon valley floor is now 1.5m higher than its present-day confluence with the River Frome.[43]

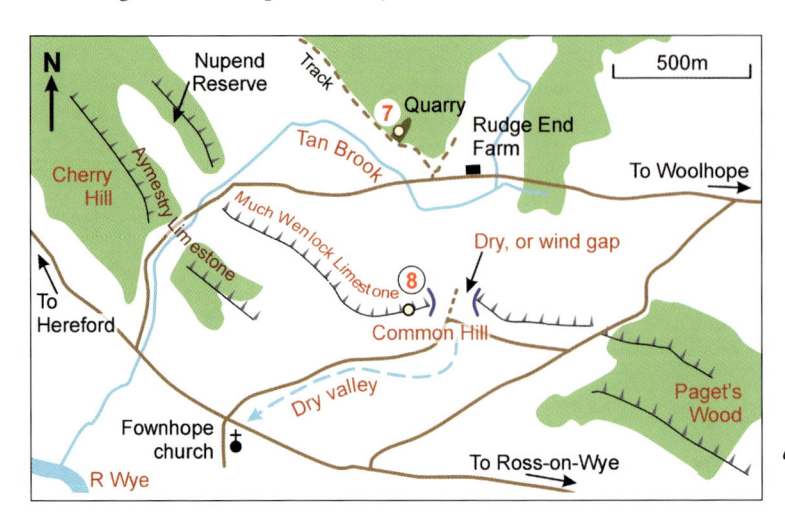

Figure 9.34 River capture near Fownhope. (© H&WEHT)

Calcareous Tufa

In contrast to the other Herefordshire rocks described in this book, one class of rock, calcareous tufa, is being formed at the present time. It consists of calcium carbonate which is deposited from solution as springs emerge from limestone rock.

In the Wye Gorge, at the Biblins, lime-saturated water emerges from Dropping Well spring (SO 552 145) (Fig. 9.35). In the increased temperature and reduced pressure in the open air, calcareous tufa is precipitated. The irregular deposits of tufa coat the cliffs of Carboniferous Lower Dolomite, now renamed the Barry Harbour Limestone. The tufa is full of holes and encloses pieces of vegetation and small animal fragments.[44]

A similar deposit of calcareous tufa is seen near Bodenham under a small waterfall where lime-saturated water emerges from the Chapel Point Limestone (Bishop's Frome Limestone) (SO 518 514). This is by the roadside on the minor road west from Bodenham past the former gravel pits and before the pub and the railway tunnel. Tufa was often used by the Normans as a building stone for churches. It is easy to cut when wet and then hardens as it dries. It can be seen at Moccas church (SO 357 433) (see p.199) and in smaller quantities in Bredwardine and Bodenham churches.[45]

Figure 9.35 Calcareous tufa seen at the Biblins. (© John Stocks)

The River Lugg flood defences

A flood alleviation scheme on the River Lugg has been designed to be in harmony with the fluvial processes and to prevent all but the most severe floods.[46] Embankments have been built, using spoil dug from the area between the banks and the river. In some places the embankments are set well back from the river to allow meander migration to continue. On some active

Figure 9.36 Lugg flood defence barrier; the 'Stank' near Mordiford. (© H&WEHT)

bends, willows were planted to stabilise the banks. Low stone weirs have been built. These reduce the velocity of floodwater by creating still water areas upstream. Many of the weirs are of uncemented blocks and these have created landscape features which can be colonised by wildlife. Downstream of Moreton-on-Lugg (SO 51 46), floodgates in the embankment, known locally as the 'Stank', allow any flood water which does overtop the banks to return to the river when water levels drop (Fig. 9.36).[47]

Herefordshire's landscape is greatly influenced by the underlying rock structure. It has been shaped by a variety of natural processes most of which are still active today. In addition, it has been adapted by human activity over the last few thousand years. There is so much to explore and enjoy in this beautiful county.

Further reading

Dorling, P. *The Lugg Valley Herefordshire: Archaeology, Landscape Change and Conservation*, Herefordshire Studies in Archaeology 4 (2007)

Dury, G.H. *The Face of the Earth*, Penguin (1977)

Harris, C. *Periglacial Landforms* in Stephens, N. (Editor), *Natural Landscapes of Britain from the Air*, P142 CUP Archive (1990)

Hey, R.W. 'The Pleistocene history of the Malvern Hills and adjacent areas', *Proceedings of the Cotteswold Naturalists' Field Club,* 33 (4) (1961), pp.185-191

Hey, R.W. 'High level gravels in and near the Lower Severn Valley', *Geological Magazine*, vol 95 (2) (1958), pp.161-168

Hey, R.W. 'Pleistocene deposits on the west side of the Malvern Hills', *Geological Magazine*, 96 (1959), pp.403-417

Hey, R.W. 'Pleistocene gravels of the Lower Wye Valley', *Geological Journal*, 26 (1991), pp.123-136

Hey, R.W. *The lower Wye valley: Geomorphology and Pleistocene geology*, in Lewis, S.G. and Maddy, D., *Quaternary of the south Midlands and Welsh Marches: Field guide*, London: Quaternary Research Association (1997), pp.61-62

Hilton, K. *Understanding Landforms*, (General Editors H. Tolley and K. Orrell), Macmillan Geography (1986) pp.68 & 115

Monkhouse, F.J. *Principles of Physical Geography*, Hodder and Stoughton (1975)

Pocock, T.I. Terraces and drifts of the Welsh border and their relation to the drift of the English Midlands. *Zeitschrift für Gletscherkunde, für Eiszeitforschung und Geschichte des Klimas*, 14 (1925), pp.10-38

Richards, A.E. 'Middle Pleistocene Glaciation in Herefordshire', *Transactions of the Woolhope Naturalists' Field Club,* 51 (2003), pp.86-96

Stamp, L.D. *The Structure and Scenery of Britain*, Harper Collins New Naturalist Series, no. 4 (1946)

Trueman, A.E. *Geology and Scenery in England and Wales*, Revised by J.B. Whittow, and I.R. Hardy, Penguin Books (1971)

Also see the Geological Conservation Review, vol. 13 (1997) *Fluvial Geomorphology of Great Britain*, Gregory, K.J. (Ed.)

10 The Impact of Geology on Herefordshire and its People

Geological processes not only underpin the natural landscape of Herefordshire but also have extensively influenced – and continue to do so – the human-made landscape and indeed the fortunes and way of life of many of the county's inhabitants. Herefordshire is a county almost without coal, with only minor igneous intrusions and, other than a source of agricultural lime, without major mineral resources. It is a county without geological raw materials to be economically exploited directly for wealth creation, though with soil well suited to growing agricultural crops. The later stages of the industrial revolution largely bypassed Herefordshire, which was outstripped by counties richer in mineral resources, and until fairly recently the greater part of Herefordshire's population has earned its living from agriculture and its allied industries.

Minerals

Before the secrets of metalworking were discovered, found materials provided early humans with vital resources. Simple hand axes made from quartzite pebbles from the conglomerates of the Sherwood Sandstone group (late Permian and Triassic Periods) are occasionally found, heavily eroded and re-deposited *c.*150,000 years ago in sands and gravels deposited by the ancient Mathon River (Fig. 9.11). Flint, which is not native to Herefordshire but was transported into the county by human and glacial movements, provided material for stone-age people to knap weapons and tools. Evidence in the form of these artefacts, and of flint cores and flakes, can be found in places throughout Herefordshire, sometimes concentrated in what appear to have been manufacturing sites (Fig. 10.1).

Figure 10.1 Flint implements from Chase Hill, Ross-on-Wye. (© Hereford Museum)

By the Iron Age, iron-based metallurgy had become an important part of the region's economy, exploiting the limonitic iron ore from the Weston-under-Penyard district and the Carboniferous ore of the Forest of Dean. In Roman times iron production from the Forest of Dean played a significant part in supporting Roman garrisons as far away as Cirencester. Iron was made using the bloomery process which created a hot spongy mass (a bloom) of soft but not liquid iron, mixed with glass from impurities, which could be worked into the required shape (Fig. 10.2).

The production of iron continued through the Middle Ages using the bloomery process but, by the 16th century and well into the 19th, water-powered bellows enabled blast furnaces to operate effectively and forge-hammers to work the blooms into wrought iron. By 1590, if not a few years earlier, an ironworks had been established at Bringewood in the Downton Gorge (Bringewood Forge; SO 4536 7496) which was leased to Richard Knight in 1698.[1] Later acquiring the freehold, Knight developed the works into a thriving business. Iron ore was quarried from deposits on Clee Hill, at that time a detached portion of Herefordshire, whilst the heavily wooded sides of the gorge and the surrounding estates provided coppiced timber for converting into charcoal for fuel. Limestone could be quarried locally, while the rapid flow of the River Teme through the gorge provided water power for blast furnace bellows and hammers.

Figure 10.2 A large Roman anvil made from a single bloom and a testimony to the skill and complexity of the Roman iron industry. (© Hereford Museum)

Similarly, the Forest of Dean had all the resources needed for iron-working: plenty of wood for charcoal, limonitic iron which did not require a flux, limestone which could be imported and a good water supply controlled by weirs to drive water wheels which worked the bellows and hammers. Iron working took place at New Weir Forge, on the right bank of the River Wye, just downstream from Symonds Yat West.

Coal has been found on Howle Hill near Ross-on-Wye, an outlier of the Forest of Dean coalfield, where there was for a few years in the 1970s an opencast mine (SO 610 204). Following the publication of William Smith's first table of strata there were speculative and costly ventures to source coal across the country and from 1809 something akin to the gold rushes in California and elsewhere broke out in Herefordshire and neighbouring Radnorshire. On the Street Court Estate near Leominster, Mrs Marianne Atherton and her adviser sought advice from William Smith and in January 1811 sent him specimens from their borings for coal. By 1816, after more borings but no coal, Marianne Atherton was severely in debt and subjected to civil actions to repay her creditors.[2] It is surprising that more than a hundred years later, after the geological basics had long been well established, expensive explorations for coal were still being pursued in north-west Herefordshire (at the aptly named 'Folly Farm').[3]

Other fruitless mining operations have included the search for gold. In the 18th century a pit 220 feet deep was dug on the Malvern Hills at a place still called 'the Gold Mine' (SO 7690 4416). It seems that the gold-coloured mica flakes which can still be found there in the soil were mistaken for the metal. Nevertheless, gold does exist in the Malvern Hills, in the alluvial deposits of local streams in the south Malverns.[4]

South Herefordshire has experienced its own small 'Gold Rush'. Old shafts and trenches within the Huntsham Hill Conglomerate in Penyard Park Hill bear testament to some abortive 19th-century investment following on the discovery of small and uneconomic deposits in the Forest of Dean.[5] A gold mine – or at least prospecting shafts and trenches – in Lea Bailey Inclosure (SO 635 197) near Weston-under-Penyard has long been abandoned for gold production and indeed it is doubtful whether any gold was actually mined. Flakes derived from the micaceous band within the Tintern Sandstone in Penyard Park (SO 620 225) were once quarried, possibly for use as a paint additive.

Peat, which was used as fuel, was at one time widespread in the county but modern agricultural practices, in particular drainage, have destroyed almost all deposits.

During the Middle Ages, and until about 1620, wool was the main source of wealth for the city of Hereford and the county's market towns, with spectacular churches built in the towns of Leominster, Ledbury and Ludlow in Shropshire. Cloth making involves 'fulling', or 'waulking' ('walking') which removes the lanolin from the wool and also cleans, felts and compacts the cloth. Stale urine (known in Herefordshire as 'sig') is an effective medium for this process as is Fuller's earth and during the Middle Ages they both seem to have been used together. The name of a house with its outbuildings, Walk Mill (SO 659 552) near the River Frome at Bromyard, betrays the buildings' origins. Murchison noted that Fuller's earth under the name 'Walker's earth', or 'soap' was sometimes used by the country people for cleansing purposes.[6]

Fuller's earth (bentonite or more correctly calcium montmorillonite) is a clay deposit formed from volcanic ash which has started to break down into its constituent clay minerals. During the Silurian Period, distant volcanic activity created many such layers – 'bentonite bands' – and a metre-thick band is exposed in Mortimer Forest close to both Ludlow with its fulling mills and to the village of Pipe Aston with its local industry of tobacco pipe making. At Pipe Aston a 17th-century clay pipe kiln has been discovered (SO 468 719).[7] In later years, clay was transported to Pipe Aston from Broseley in Shropshire, the main centre of clay pipe manufacture in the region, presumably first reaching Broseley by Severn trow from the Cornish china clay mines.

More information on the topics in this section may be found in Keith Ray's *The Archaeology of Herefordshire: An Exploration* (2015). These topics and the history of the local geological societies are discussed also in K. Andrew's 'History of Geology in Herefordshire' (to be published).

Rock
Limestone, principally Bishop's Frome Limestone, was quarried and burned to produce lime for mortar and lime-wash for protecting buildings, for use in the tanning of leather and for spreading on lime-deficient agricultural land. The remains of a great number of limekilns can be found in the limestone regions of the county, in the vicinity of the Woolhope Dome for instance and in the Golden Valley.[8] Again, to reduce the acidity of agricultural land and to improve its fertility, marl, with its content of lime and other minerals, has been dug from 'marl pits' in the Moor Cliffs (Raglan Mudstone) and Freshwater West (St Maughans) Formations and spread, untreated, on fields.

Clay, mostly from mudstones of the Lower Old Red Sandstone, is a geological resource which has been widely put to use in Herefordshire. A Romano-British pottery kiln has been unearthed near Stoke Prior (SO 535 566), whilst craftsmen with their medieval kilns and, later, local brick, tile and pipe makers exploited this resource on a larger scale. One such business existed until the late 20th century in Linton (SO 669 541), digging out clay from the mudstones of the Freshwater West Formation to manufacture products which found their way into the county's buildings. For technical reasons and competitive pressures, this once thriving activity in Herefordshire has died out, like other mineral-based industries in the county.

Sandstone roofing tiles are quarried from a 'delph' ('delf' derived from 'delve', an excavation or pit, as well as to dig) and require a fissile rock of particularly good quality. A suitable material, much used in Kington, was quarried from sandstone of the Downton Castle Sandstone Formation on Bradnor Hill (SO 291 577), and at Abbey Dore the church has recently been re-roofed with tiles from two former delphs which were reopened for that purpose.

Fine-grained sandstone of an as yet unknown origin was exploited by medieval carvers, notably in examples of the 12th-century Herefordshire School of Sculpture to be seen in Kilpeck church and in the fonts of the churches at Castle Frome and Eardisley.

The most evident of the county's geological resources are the many varieties of building stones which can be found in the county. Detailed information is beyond the scope of this chapter and can be found elsewhere.[9] Stone extracted from countless small quarries for local use provided the material, along with timber framing, for most of the county's buildings before the adoption of brick and later methods of construction. These later techniques may appear to use stone traditionally as a load-bearing material when, in fact, it is used solely for the sake of appearances to face a building which may be of metal frame and concrete construction. Also, where repairs or additions to existing buildings have been involved, use has often necessarily been made of similar but not identical stone from outside the county; recent restoration of the tower of Hereford Cathedral, which is basically of Devonian sandstone construction, has used Triassic sandstone from quarries in Shropshire. But this has not been a one-way traffic. For instance, sandstone was exported from the county for the 19th-century restoration of Monmouth's parish church. At present there are four significant, though small, sites in Herefordshire quarrying sandstone of the Freshwater West (St Maughans) Formation for building purposes – one near Bredwardine, at Harewood End and two near Pontrilas.

Devonian sandstones and siltstones together with, to a far lesser extent, Silurian and Carboniferous limestones are the predominant building stones which have been used in Herefordshire, and it is in the county's old churches that accessible examples of these can best be seen. Devonian sandstones vary in quality and siltstones sometimes make very inferior building stone which, when used in exterior rubble walls, suffer badly from erosion. A more durable and stronger material, such as limestone, was often favoured for exposed quoins and lintels. Tufa, a light but strong stone, has been used, mostly in areas close to deposits, such as in the church at Moccas (Figs. 10.3 and 10.4). In Mathon (Fig. 10.5) and Colwall churches (Fig. 10.6) a variety of building stones, including Precambrian rocks and glacial

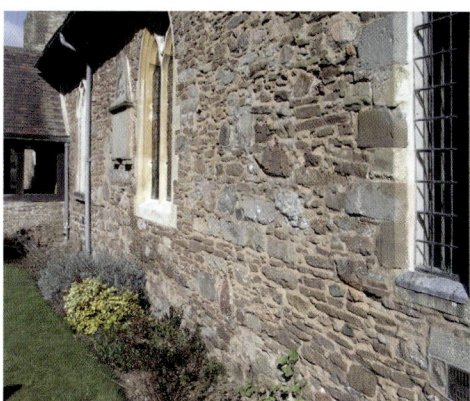

Figure 10.3 (above) The bellcote of Moccas Church (SO 357 431), built of tufa, with two sandstone gravestones in the foreground. (© John Stocks)

Figure 10.4 (top right) Tufa rubble walling with two blocks of sandstone in Moccas church tower. (© Charles Hopkinson)

Figure 10.5 (middle right) A variety of rocks set in the south wall of the nave of Mathon Church (SO 734 458). (© Charles Hopkinson)

Figure 10.6 (right) Local stones used in Colwall Church (SO 7390 4231). (© John Payne)

erratics, have been incorporated into the mostly rubble walls of the buildings. Triassic Bromsgrove Sandstone, relatively bright red compared to Devonian sandstone, has been used in the 19th-century restoration of Whitbourne Church (Fig. 10.7). Abbey Dore church is built mostly of Old Red Sandstone but has quoins of harder Devonian gritstones and conglomerates. Although there are Precambrian rocks on the Malvern Hills, because of their hardness, and particularly because they are highly fractured, they cannot be dressed for building purposes and do not feature to any great extent in Herefordshire's buildings. There is only a single major old quarry (Gardiner's Quarry) of the Malvern rock within the county's boundary. Stone of the Freshwater West Formation from Cradley Quarry (SO 718 475) has been used in several major buildings in Malvern (e.g. Malvern College chapel and what is now the Council House).

Figure 10.7 The Devonian sandstone tower of Whitbourne Church (SO 726 569) with its west window which has been rebuilt in Triassic Helsby Sandstone (Bromsgrove Sandstone). (© Charles Hopkinson)

At Goodrich Castle (Fig. 10.8), the mid-12th-century keep is built of grey-green sandstones from the Lower Old Red Sandstone and blocks of local quartz conglomerate. For the late 13th-century reconstruction of the defences, sandstone from the Brownstones Formation with traces of conglomerate (later in the Old Red Sandstone) was all, or largely, quarried on site to form the castle's moat. One cannot but admire the skill of medieval masons who, in this instance, developed rectangular stone platforms on dipping strata in the ditch on which to build the massive cylindrical south-eastern and south-western mural towers which are still standing 700 years later.

The once abundant building stone quarries in the county have mostly been abandoned but the production of aggregate (crushed rock, largely for roadstone for which the Old Red Sandstone is unsuitable) continues. It is presently quarried from the Silurian limestone at Leinthall Earls (SO 443 685; Fig. 2.15) in the north of the county and at Perton Quarry (SO 595 398) in the Woolhope Dome. At Leinthall Earls up to 2,000 tons of aggregate are produced each day.[10] Precambrian rocks of the Malverns Complex were extensively quarried for the manufacture of aggregate at, for example, Tank Quarry (SO 768 470); Hollybush and Gullet Quarries ceased production as late as the 1970s, after supplying much stone for the construction of local motorways.

Figure 10.8 The mid-12th century, grey-green sandstone keep of Goodrich Castle (SO 577 200), contrasting with the late-13th century towers and curtain wall of reddish sandstone.
(© John Stocks)

Older large-scale maps of Herefordshire, as well as evidence on the ground, indicate the many sand and gravel pits that once existed in the county and which have been abandoned. Quaternary deposits of sand and gravel from the river terraces and glacial moraines in the river valley of the Lugg continue to be extracted at Wellington (SO 507 475) and elsewhere for the production of aggregate which is used, among other things, in the manufacture of concrete. Glaciofluvial gravel deposits from the Wye glacier have been exploited in the past at Stretton Sugwas west of Hereford (SO 455 422). In the valley of the River Arrow, north of Pembridge, boreholes have indicated massive deposits of gravel, but gravel and sand deposits in the east of the county have not been widely exploited in view of the high water table. The landscaping and establishment of a nature reserve in and around the old gravel pits at Bodenham (SO 525 512) is an example of how the restoration of a geologically exploited landscape can enhance the local flora and fauna. The sand and gravel deposits of the Mathon River from the pre-Anglian period have been extracted from several workings south of Cradley.[11]

Settlements

Geology considerably influenced Herefordshire's pattern of settlements which were to grow into the county's towns and villages. Natural caves, such as King Arthur's Cave (SO 546 156; Fig. 2.25) and other rock shelters in the Carboniferous Limestone near Whitchurch, occur at the top of the gorge some 100m above the present level of the River Wye. They were formed when the river occupied an earlier, higher-level course and they provided shelter for Upper Palaeolithic communities (*c*.40,000-10,000 years ago). The builders of King Arthur's Stone (SO 319 431; Fig. 2.18) – a later Neolithic chamber tomb near Dorstone in the Golden Valley – used an enormous though now broken sandstone capstone weighing some 25 tonnes. Later still, Iron Age hillforts, which abound in Herefordshire, made use of features of the landscape to establish viable communities protected by massive artificial earthworks and sophisticated gateway defences, whilst evidence has been found not only of such communities' agricultural activities but also of scattered farmsteads on lower lands.

Among important considerations in choosing places for settlement would have been: dry sites near to natural supplies of water; proximity to fords and bridgeable stretches of rivers which could be used for trade and water mills; and proximity to suitable land for agriculture. Early cross-country ridgeways to facilitate communication and trade were developed on higher, drier ground, away from forests and marshes. One such ridgeway ran along the high land to the east of the Golden Valley (now partly through private property) from Merbach Hill (SO 304 447), by way of Arthur's Stone, Upper Bodcott and through Moccas Park to the Iron Age hillfort in Timberline Wood (SO 388 369). Medieval castles such as Wigmore Castle were sited to make, with additional human-made artifice, the most effective use of the natural terrain, and this appreciation of the potential of the landscape can often be recognised in the location of medieval farmhouses and later country houses.

Place-names have on occasion an obvious geological context.[12] For instance, Staunton on Arrow is a 'settlement on stony ground' on the River Arrow and Clehonger a 'sloping wood on clay soil'.

Of vital importance to Herefordshire's tourist trade is its countryside; indeed, the appreciation of the British landscape, which developed in the late 18th century and is known as the Picturesque Movement, has some of its roots in the county. It led to the popularity of purely scenic tourism, particularly in the Wye gorge.[13] Examples of Picturesque art are shown in Figs. 10.9 and 2.12.

Figure 10.9 The Cliff opposite the Hay Mill by Thomas Hearne (1785). This Picturesque painting may be compared with Fig. 2.17 which shows the scene from much the same place. (Private collection. Reproduced here by permission of Prof. Charles Watkins, Nottingham University)

> Box 10.1   **Herefordshire and the Picturesque Movement**[14]
>
> The concepts involved in the Picturesque artistic movement were introduced in Boxes 2.1 and 2.5. The ideas stemmed in a major way from individuals in Herefordshire who were profoundly influenced by the local landscape, often in part created by themselves as landowners.
>
> 'The Picturesque' may be said to be an aesthetic of the natural order; an awareness of its magnificence, particularly as displayed in the landscape. In this, it can be seen as a link between the Enlightenment of the 18th century and the full flowering of the Romantic Movement in the early 19th century, containing elements of both. During this period the expansion of the scientific perspective (e.g. the geological insights of James Hutton (1726-1797), a distinctive figure in the Edinburgh Enlightenment, along with such figures as James Watt and David Hume) was matched by a growing awareness of the magnificence and aesthetic beauty of the natural order. Such 'picturesque' ideas were particularly focussed in Herefordshire and the Marches, the area on the border with Wales,[15] although by no means confined to this area. In the lower Wye area,[16] this idea was developed in *Observations on the River Wye and Several Parts of South Wales* (1782) by Gilpin particularly in relation to a growing 'tourist' interest in the natural landscape (sometimes for artistic purposes). Gilpin welcomed the industry along the river valley as adding further dramatic elements to the landscape. In the middle reaches of the Wye, with its fertile soils, river meadows, woodlands and apple orchards, the emphasis was more that of the sensitive landowner and farmer – e.g. Uvedale Price and his *Essay on the Picturesque* (1810). Further north, on the River Teme at Downton, Richard Payne Knight radically extolled the virtues of the more sublime beauty of nature's magnificence in *The Landscape: a Didactic Poem* (1795). In due course, ideas of the Picturesque would feed into the Romantic Movement, but in its origins, this aesthetic was closely associated with Herefordshire and its Palaeozoic geology.[17] In more recent times, Picturesque ideas have combined with more specifically scientific ones to contribute to modern day thoughts about ecological conservation. Natural England, for example, is now responsible for the Downton Gorge[18] and the Herefordshire Wildlife Trust has a substantial number of reserves on The Doward, the Carboniferous limestone area with the Wye Gorge in the south of the county.

Agriculture

From early times, the fertile soils of the county's lowlands, including the glacial tills of the western part of the county and the thinner ones of the uplands, have provided resources which have been successfully exploited by farmers in various agricultural systems. Geology is partly responsible for Herefordshire's relatively benign climate as, by and large, the county lies in the 'rain shadow' of the Welsh hills and the Black Mountains to the west of it, thus enjoying a lower level of cloud cover and a higher level of sunshine than it might otherwise have done. The mean annual temperature is roughly two degrees higher than in central Wales and sunshine hours about 10% greater.[19]

Figure 10.10 Herefordshire red soil. A view westward towards the Black Mountains and the upper Wye river valley. (© Jack Wilson)

The basic constituent of natural soil is the degraded result of the physical and chemical weathering of rock. This is in general the underlying material though in some areas glacial drift provides a soil unrelated to it. With the addition of humus – organic material formed from the decay of plant and animal remains – a fertile agricultural soil is established. The distinctive red soils of much of Herefordshire (Fig. 10.10) gain their colour from the underlying Old Red Sandstone of the Devonian and are chemically broadly 'neutral', but can be slightly alkaline or slightly acidic.[20] The soils derived from the limestone rocks of the Silurian and Carboniferous (for example in the Woolhope Dome and in the Wye Gorge respectively) are alkaline, whilst the soils of the Precambrian Malvern Hills are predominantly acidic.[21] It is upon the acidity or alkalinity of the soil, which can be changed by the application of lime and marl (see above), that the variety of natural vegetation largely depends.

The contrast between lime-loving and acid-loving plants, and hence evidence of the underlying geology, is plain to see in the Kington area. Here, in Hergest Croft Gardens in acidic soils derived from the underlying rock, rhododendrons, azaleas and other acid-loving plants thrive; they will not be found, except in human-made ground, in the Old Red Sandstone soils to the east. The belts of Bishop's Frome Limestone in the county with their nodules of 'cornstones' (or 'calcrete' in current nomenclature) produce an alkaline soil which favours the growth of plants such as the rock rose, scabious, orchids and carline thistle. This phenomenon can be seen in the Olchon valley when looking up at the band of a similar rock, the Ffynnon Limestone, at an altitude of about 540m OD on Black Hill ('The Cat's Back'; SO 276 344); here a clear, brighter green line of lime-loving plants is visible amid the darker heathers of the ridge. On the thin soils of the Malvern Hills, with their variety of geology, the vegetation is mainly calcifugous, thriving in acidic soils, but with some areas of calcicolous species which flourish in alkaline soils. Ash and beech trees grow particularly well in alkaline soils whilst other woodland species and those cultivated in orchards are more tolerant of, or favour, a more neutral or slightly acidic soil. The county's various nature reserves are evidence of the growing interest in conservation with their particular flora reflecting the underlying geology.[22]

Most of the major rivers of the county flow into and through the county from the west, depositing huge amounts of geological detritus in the form of silt to form 'flood meadows'. These, with their increased fertility, were traditionally used as grazing and hay meadows. It was on the flood meadows of north-west Herefordshire that the Hereford breed of cattle was developed in the late 18th and early 19th centuries. The flood meadows provided early grazing for this increasingly important breed of beef animal – in due course to be exported as breeding stock across the world.

Such was the value of the flood meadows that in the late 16th century an attempt was made by Rowland Vaughan to reproduce them in the Golden Valley.[23] Extensive evidence can still be found on the ground south of Peterchurch of the channels that were to flood the valley floor from the River Dore. Since the beginning of the Second World War, however, with the need for home-grown food and with post-war targeted incentives for agriculture, these lands have been increasingly exploited for arable crops as have other fertile lands formed from glacial deposits. Clays and marls with their water retentive properties have provided an ideal environment for orchard and soft fruit production, as well as for growing fine timber; oak was once referred to as 'the Herefordshire weed' and has provided much of the material for the construction of the county's half-timbered buildings. Woodland was one of the county's vital resources being used for many purposes, not least the creation of charcoal which was an important fuel, particularly for iron smelting, before being superseded by coal.

Transport

While the rich, clay soils of much of Herefordshire have been beneficial for agriculture they have proved in the past a hindrance to communication. During the winter months the county's roads were for centuries notorious, being often well-nigh impassable to wheeled traffic, while in the summer the alternative of river traffic could be unreliable in view of unpredictable river levels. It was not until the development in the 18th and 19th centuries of turnpikes, canals, tramways and railways, that travel and the carriage of goods within the county were transformed.

Energy

Groundwater, springs and rivers have supplied the vital resource of water. Groundwater, from precipitation soaking into the ground and underlying rocks, is a valuable resource for agriculture, business and private use. Natural springs and boreholes tap underground water in rock fractures, aquifers and subterranean reservoirs, which flows or is pumped to the surface.[24] Rivers have not only provided water but also for many centuries furnished a vital resource – water power was the primary source of energy before the Industrial Revolution and indeed a source of energy for long afterwards. A water mill at Rowden (SO 629 566) was producing animal feeds as late as *c.*1947. At Poston Mill (SO 356 371), the remains can be seen of a turbine which, unusually for Herefordshire, replaced a waterwheel, and at one time there was a turbine-driven electricity generator on the River Frome at Stretton Grandison. Modern, small hydroelectric generating systems have a role to play in the provision of electricity for local consumption.

Seismic surveys and drilling for oil and gas which have taken place in the Fownhope and Collington districts have failed to find oil or gas and as yet no hydrocarbons have been detected in Herefordshire. Hydraulic fracturing – fracking – for the recovery of shale gas may in the future take place in the county. Possible sites for exploration for shale gas have been identified by the former Department of Energy and Climate Change (DECC) in the Lower Palaeozoic shales in the Eastnor, Fownhope and Much Marcle districts.[25] At present, though, it appears that there are many more potentially profitable sites elsewhere in the United Kingdom to be exploited first.

Geothermal energy, another geological resource, occurs naturally in the earth, particularly near tectonic plate boundaries, and can find its way to the earth's surface through phenomena such as hot springs, geysers, volcanoes and fumaroles, none of which now feature in Herefordshire. However, this energy can be tapped subterraneously, for heating purposes and energy generation, using new technologies such as ground source heat pumps which will no doubt be exploited in the county in the search for sustainable and acceptable sources of energy. Herefordshire exhibits a relatively low geothermal heat flow compared to the neighbouring Worcester Basin, which is a prime area for geothermal exploitation.[26]

Geological Hazards

Geohazards are geological conditions which either threaten to give rise to damage or actually do damage the environment, thus endangering the human population. Their impact can be massive and widespread – as in a tsunami – or local in scale as has happened in the county.

Evidence, in the form of locally disturbed ground, can be found of the large landslip, 'The Wonder' (SO 633 365). This occurred in 1575 on the eastern side of the Woolhope Dome.[27] During the winters of 1946-7 and 2013-4, with their abnormal snow and rainfalls, there were landslips along a steep hillside (SO 591 507) between the villages of Pencombe and Ullingswick, in a locality prone to instability (Fig. 10.11).[28] Here, relatively strong rocks of the Freshwater West (St Maughans) Formation overlie weaker and less permeable rocks of the Moor Cliffs (Raglan Mudstone) Formation. Both landslips damaged a local road and affected up to one and a half hectares of agricultural land to the extent that mature trees were moved upright to new positions down the slope without visible damage. Sinkholes are another symptom of ground instability which can develop as water dissolves bedrock, such as limestone and chalk, leaving ground cover unsupported until it finally collapses. An example is found on Great Doward (SO 557 156), near Symonds Yat and adjacent to some old iron ore mines. As for river flooding, responsibility for this can be laid at the doors of both geology and the weather.

Figure 10.11 The road between Pencombe and Ullingswick damaged by the landslip of 7 February 2014. The ranging pole is marked in half-metre lengths. (© Charles Hopkinson)

The Hereford region is one of the most seismically active parts of the UK. Three significant tremors occurred in the late 19th century and several smaller ones more recently. That in 1896, with an estimated magnitude of 5.3 on the Richter scale, caused minor structural damage in the county; all seven chimney-stacks were shattered at Hereford railway station.[29]

Radon is a natural, dangerous radioactive gas usually emanating from granitic rocks, but is not the problem in Herefordshire that it can be elsewhere, for instance, in Devon and Cornwall. Within the county only in the Kington district does radon appear to present any significant risk.[30] Stricter than normal building regulations to protect buildings against radon are also enforced in the Malvern Hills locality where there is a higher concentration of the gas than elsewhere. Another potentially dangerous gas, methane which is usually associated with landfill sites and organic features such as peat bogs, does not seem to present a significant risk in the county.

Conclusion

No one knows when or how in the future geology will shape Herefordshire's landscape and the lives of its inhabitants. Looking at Great Britain as a whole – indeed Europe and the world as a whole – there is no reason to doubt that sooner or later geology will once again affect landscape and climate, drastically influencing people's way of life and countries' economies. Some 2.6 million years ago the rising sea floor between what we now know as South and North America formed a bridge between the two landmasses. It is thought that the effect that this had on ocean currents in the Atlantic was one of the mechanisms which led, to a greater or lesser degree – opinions differ – to an increase in ice in the northern hemisphere and the development of an ice age.[31] At present, we live in an interglacial but the relatively benign environment we enjoy will not last for ever!

Further reading

British Regional Geology; The Welsh Borderland (3rd edition), Institute of Geological Sciences (now the British Geological Survey) (1971) 118pp.

Memoirs, Sheet Explanations and Sheet Descriptions by the British Geological Survey. (See pages *xix-xx* on 'EHT and BGS Publications'.)

For information on mineral resources, see, for example:
   The BGS Memoirs (as above)
   An Inventory of the Historical Monuments in Herefordshire, Royal Commission on Historical Monuments England (1931-4), 3 vols
   Mineral resources Information for Development Plans, Herefordshire and Worcestershire; Resources and Constraints, Report no. WF/99/04, British Geological Survey (1999), 37pp.

Chapman, S.D. and Chambers, J.D. *The Beginning of Industrial Britain*, University Tutorial Press (1970)

Ray, K. 2015 *The Archaeology of Herefordshire: An Exploration*, Logaston Press (2015)

Websites
   Herefordshire Council, 'Herefordshire Through Time': htt.herefordshire.gov.uk/
   Herefordshire and Worcestershire Earth Heritage Trust: www.earthheritagetrust.org/
   Abberley and Malvern Hills Geopark: www.geopark.org.uk/
   Herefordshire Botanical Society: http://ralph.cs.cf.ac.uk/HBS/HBS.html
   Herefordshire Wildlife Trust: www.herefordshirewt.org/
   Malvern Hills Conservators: www.malvernhills.org.uk//

Appendix The 'Hereford Speech' by John Masefield, Poet Laureate

Words spoken to the Right Worshipful, the Mayor, the Councillors and Aldermen of Hereford, on Thursday, October 23rd, 1930:

I have now to thank you for the great and beautiful honour that you have paid me in giving me the Freedom of this City.

It is a very great honour to be received into any city of men and women, as a fellow citizen, with privileges that few of the citizens enjoy.

I am the more conscious of the honour, since you pay it to me because I am a poet. Often a poet is a solitary, who is not at one with his community, and only enters it to wound its members and himself, to rebel against it and outrage it because of something in his mind that is not in this world at all; and cannot adjust itself, but wants the moon or some image of the moon, and so lives restless and dies wretched, leaving behind the images of his wants.

Many poets would say 'I am nothing, and belong nowhere, but in my mood I go into a place that is better than anywhere, and the king and queen of that place are greater than anybody here and give me words to say.'

I believe that life is an expression of some Law, or Will, that has a purpose in each of its manifestations. I believe that this world is a shadow of another world.

And looking intently on what is brightest and most generous in this world, which is but a little and dim thing compared with the real world, the great beauty and bounty and majesty of the real world are borne in upon the soul.

I am linked to this County by subtle ties, deeper than I can explain: they are ties of beauty. Whenever I think of Paradise, I think of parts of this County. Whenever I think of a perfect Human State, I think of parts of this County. Whenever I think of the bounty and beauty of God, I think of parts of this County.

I know no land more full of bounty and beauty than this red land, so good for corn and hops and roses. I am glad to have lived in a County where nearly every one lived on and by the land, singing as they carried the harvest home, and taking such pride in the horses, and in the great cattle, and in the cider trees. It will be a happy day for England when she realises that those things and the men who care for them are the real wealth of a land: the beauty and bounty of earth being the shadow of Heaven.

Formerly, when men lived in the beauty and bounty of earth, the reality of Heaven was very near; every brook and grove and hill was holy, and men out of their beauty and bounty built shrines so lovely that the spirits which inhabit Heaven came down and dwelt in them and were companions to men and women, and men listened to divine speech. All up and down this County are those lovely shrines, all of the old time.

I was born in this County, where there are so many of those shrines, the still living evidence that men can enter Paradise. I passed my childhood looking out on these red ploughlands and woodland pasture and lovely brooks, knowing that Paradise is just behind them. I have passed long years thinking on them, hoping that by the miracle of poetry the thought of them would get me into Paradise, so that I might tell people of Paradise, in the words learned there, and that people would then know and be happy.

I haven't done that, of course, or begun to, but in giving me this freedom you recognise that I have tried, and I therefore thank you.

Acknowledgement is gratefully given to the Society of Authors as the Literary Representative of the Estate of John Masefield to quote from his 1930 'Hereford Speech'.

Glossary

Acadian Orogeny	The Acadian Orogeny refers to a late phase of the main Caledonian orogeny, which occurred in early Devonian times (400Ma). This folding event was after the formation of the Lower Old Red Sandstone and is indicated within Herefordshire by the absence of the Middle Old Red Sandstone.
Airfall tuff	A fine-grained deposit formed by the fall out of ash-grade material from the air during an eruption and which is subsequently lithified.
Anorthite	The greyish calcium mineral of the plagioclase feldspar group. Its chemical formula is $CaAl_2Si_2O_8$.
Arenaceous	A term applied to sediments or sedimentary rocks composed wholly or partly of sand-sized fragments.
Argillaceous	A term applied to sediments or sedimentary rocks composed wholly or partly of clay or silt-sized fragments.
Asturian	A phase of deformation within the Variscan Orogeny; 300 to 280Ma
Back-arc basin	A sedimentary basin situated on the upper plate within a subduction zone and behind the volcanic arc. The crust here is usually undergoing extension due to rising melts from the subducting slab, leading to localised volcanism (e.g. Sea of Japan).
Ballstone reef	Flat-bottomed lenticular mass of unstratified limestone composed of large colonies of both corals and stromatoporoids set in a calcite mudstone matrix derived from the decomposition of calcareous algal material. Synonymous with 'patch reef'.
Bentonite clay	A calcareous mud rock composed almost totally of the clay mineral montmorillonite and colloidal silica produced by the alteration of volcanic debris usually in the form of tuffs with glassy rather than crystalline particles. Palaeozoic bentonites, such as those in the Silurian of Herefordshire, are now altered to illite-smectite clays. They are the result of major volcanic eruptions and may be deposited over a large area. They can therefore act as useful stratigraphic marker horizons.
Biosparite	A limestone made up of both skeletal remains of calcareous organisms and clear calcite (spar) crystals.
Bioturbation	The reworking of sediment layers by the feeding and burrowing activities of organisms usually worms or crustacea.
Boudinage	A tectonic stretching process whereby an individual rock layer is first thinned and then ruptured into separate sausage-like shapes called boudins.
Braided river	A river carrying an excessive sediment load and possessing an intricate network of dividing and rejoining channels. Braiding occurs when sand and gravel bars are deposited within channels.

Breccia	A coarse-grained sedimentary rock composed of angular fragments set within a mineral cement or a fine-grained matrix.
Bryozoa or Bryozoan	A group of colonial aquatic invertebrates characterised by a branching structure and composed usually of calcium carbonate. They are important reef builders.
Cadomian Orogeny	An extensive mountain building phase lasting from approximately 700 to 450Ma recorded in a belt running from the northern Appalachians (USA) to Armorica (Brittany) and the southern British Isles.
Calc-alkaline	The composition of the basalt-andesite-rhyolite suite of igneous rocks that is typically formed in island arcs and subduction zones at continental margins. Characterised by plagioclase feldspars (rich in calcium and sodium) with or without orthoclase feldspar (rich in potassium).
Calcrete palaeosols	These are formed by an accumulation of calcium carbonate within the soil itself. This is caused by some downward movement of lime-rich water but by far the most important process is the upward movement by calcium-rich groundwater through capillary action. This is aided by high ground surface temperatures. Examples, such as the Chapel Point Limestone, occur in the Lower Old Red Sandstone.
Caledonian (Orogeny) (adj. Caledonoid)	A series of orogenic episodes affecting the north and west of the British Isles during Lower Palaeozoic time which reached their peak with the collision of the former continents of Laurentia, Baltica and Avalonia in late Silurian times. The structures resulting from these mountain building events are said to have a Caledonoid trend, namely NE-SW across most of the British Isles.
Chlorite	A group of usually green platy minerals related to micas and commonly occurring in low temperature metamorphic rocks.
Clast	A separate fragment of a larger rock mass, of any size from boulder to clay particle, removed by the physical disintegration of that rock mass and subsequently incorporated into a sediment or sedimentary rock.
Conglomerate	A sedimentary rock consisting of more or less rounded fragments of pre-existing rocks over 2mm in diameter (granules, pebbles, cobbles or boulders) set in a finer-grained matrix of sand or silt.
Consequent stream	A stream within an overall drainage pattern whose course is determined by the overall slope of the land. In scarp and vale landscapes, consequent streams flow in the same direction as the overall dip of the strata.
Coquina	A detrital limestone made up almost entirely of sorted and cemented fossil debris. It is essentially a lithified shell bank where all of the constituents have been selectively transported by currents.
Cornstone	A concretionary, often discontinuous, limestone developed in the soil due to capillary action under semi-arid conditions (for example, within the Lower Old Red Sandstone in Herefordshire) (see also Calcrete).
Crinoid	A major group of echinoderms, popularly called 'sea lilies', as they have a long flexible stem made up of many separate sections, called 'ossicles', topped by a cup and feathery-looking branches. They also have root structures bonding them to the lagoon or sea floor. Separated 'ossicles' are common as fossils and are a major component of crinoidal limestone.
Dip	See Fig. 3.14.

Dolomite	A common white or yellow carbonate mineral, $CaMg(CO_3)_2$. It is also used to refer to the sedimentary rock composed mainly of the mineral dolomite.
Drag fold	A fold that develops in rock strata adjacent to a fault or thrust plane during or just before movement.
Escarpment	A more or less continuous cliff or slope formed at the top of the dip-slope by erosion through the thickness of a sedimentary horizon and sloping steeply in the opposite direction to the local dip-slope.
Ferromagnesian minerals	Rock-forming silicate minerals that contain iron and magnesium or both (e.g. hornblende, augite).
Fluvioglacial gravels	Part of the outwash deposits laid down by glacial melt-water streams. The sediment is often crudely sorted and stratified by this water transport.
Glaciation	A process in which the Earth's surface is covered and altered by both glaciers and ice sheets during an Ice Age. Often used as a noun to indicate a period of glaciation. 'Glacial' describes features or processes that take place during a glaciation e.g. glacial lake, glacial erosion.
Graben	The downthrown block between two nearly parallel inward-facing normal faults. The upstanding crustal blocks on each side are referred to as horsts (see separate definition).
Graptolite	Abundant in Ordovician and Silurian seas, graptolites are an extinct group of colonial invertebrates. The true graptolites were swept around the world's oceans by currents (pelagic lifestyle) and are made up of one to eight 'arms' all coming from a central point. Each 'arm' looks like a very small fret saw blade with tiny serrations each of which houses a single polyp. Fossil graptolites are the carbonised remains of these skeletons. Dendroid graptolites were the earliest graptolites to evolve and, unlike true graptolites, have many 'arms' (up to fifty in some cases) and lived attached to the sea bed. Their bell-shaped structure allowed food-bearing currents to pass through the sessile skeleton.
Greywacke	A hard, coarse-grained sandstone, usually dark grey in colour, characterised by angular particles of both quartz and feldspar together with rock clasts embedded in a clay matrix that forms more than 15% of the rock. Greywackes often exhibit fining-upwards cycles in outcrop.
Grit	A sedimentary rock consisting of medium- to coarse-grained sand. The grains are often clearly visible to the naked eye. Secondary quartz cement within the pore spaces produces a rock which is tough and resistant to weathering.
Hardness	The resistance of a mineral to abrasion measured by the Mohs Scale, a scale of hardness devised in 1822 by mineralogist Freidrich Mohs. The ten minerals selected by Mohs merely denote an order of hardness (resistance to scratching) and have no quantitative significance.
Horst	A raised block usually caused by crustal extension whose long sides are bounded by steeply inclined faults that generally dip away from each other.
Intraformational conglomerate	A conglomerate formed by the reworking of contemporary sediments such as the fragmentary mud clasts incorporated into overbank flood deposits in the Lower Old Red Sandstone of the Welsh Borders.

Lag	A deposit found on the inside of river meanders and formed of the coarse-grained sediment left behind as the finer material is swept downstream.
Laurussia	A super-continent formed by the collision of Laurentia, Baltica and Avalonia in the Silurian Period.
Malverns Complex	A calc-alkaline igneous complex which forms the Malvern Hills. It is composed mainly of diorite and granite and was intruded between 680 and 670Ma during a period of island arc magmatism.
Mantle plume	A stationary column of magma rising up in some cases from the core-mantle boundary which initiates 'hot spots' of increased volcanism often, but not always, independent of plate tectonics (e.g. Hawaii).
Meander	Many rivers show curves or loops in their stream channels. Most meandering rivers have a fairly low bed load of carried sediment and steeper banks than braided streams. Meanders tend to become more sinuous the lower the gradient of the flood plain.
Moraine	An accumulation of rock material that has been carried or deposited by a glacier. It ranges from boulders to the finest rock flour in size and shows no bedding or sorting.
Orogeny	An episode of mountain building usually involving processes of folding and associated thrust faulting, e.g. the Variscan orogeny at the end of the Carboniferous Period.
Palaeocurrent	An ancient current of water or wind. Its direction can be inferred from various sedimentary structures such as cross-bedding or ripple marks and by other methods such as the analysis of the dune types.
Palaeogeography	The study of the rocks and fossils of a specified time in the geological past can allow the drawing up of a palaeogeographic map for that time which includes the distribution of land and sea, the depth of those seas, and the overall geomorphology. This is usually superimposed on a map for the present day, for ease of understanding, so the palaeo-distances (e.g. width of shallow seas) are not accurate.
Phylum (plural: Phyla)	The largest category used to classify organisms in both the animal and plant kingdoms. It represents an assemblage of organisms having a common structural plan, e.g. Chordata (all animals with a backbone).
Pillow lava	Characterised by a mass of rounded and slightly flattened lava bodies resembling pillows. They are formed usually as basalt lavas are extruded into water. The water chills the outside of globules of lava, forming a crust, before more lava bursts forth forming more pillow shapes.
Plate	An individual slab of lithosphere which rests on and moves over the static asthenosphere.
Plinian-style eruption	A powerful volcanic eruption, named after the AD79 eruption of Vesuvius documented by Pliny the Younger, which is characterised by a large column of pyroclastic material ejected high into the atmosphere. This eruptive column spreads out into a mushroom shape when it interacts with the stratosphere.
Pluton	An intrusive igneous body that forms when magma cools and crystallises within the Earth's crust.
Proglacial lake	A lake that sits immediately in front of or just outside the limits of an ice sheet, usually fed by meltwater streams.

Pyroclastic Flow	A hot, particulate, mass of pyroclastic material moving as a very rapid surface flow away from a volcanic vent. Flow is controlled by gravity but may be partially fluidised.
Reef	A major mound-like body found only in marine environments and constructed by calcareous algae and other sessile calcareous animals, predominantly corals, which are important geologically. Patch reefs are small isolated mounds which develop on the continental shelf and are typical of the mid-Silurian Much Wenlock Limestone environment. 'Bioherm' is a synonym for any type of reef structure.
Rheic Ocean	The ocean between the African and European plates in Upper Palaeozoic times. Its closure initiated an extended period of crustal deformation (the Variscan orogeny of the late Carboniferous and early Permian (Figs. 3.10 and 3.11).
Rift valley	A major elongate depression bounded by steeply dipping normal faults resulting from crustal extension processes. Synonymous with 'graben' (see above).
Rudaceous	A term applied to sediments or sedimentary rocks composed of a significant proportion of fragments larger than sand, such as pebbles or gravel.
Shear zone	An area of ductile deformation resulting from forces acting parallel to each other but in opposite directions. The result is a displacement of adjacent layers along closely spaced planes.
Slickenside	A rock surface that has become polished and striated from the grinding or sliding motion of an adjacent rock mass which is often found along fault planes. The striations (scratch marks) may indicate the direction of relative motion.
Spilite (adj. spilitic)	An altered basalt, usually pillow lava, in which new minerals have formed caused by interchange with heated sea water wherein one mineral replaces another.
Stratum (plural 'strata')	A layer of sedimentary rock. Individual layers of rock are also referred to as beds.
Strike direction	The compass direction of a line drawn horizontally on an inclined bedding plane. The dip direction is the angle of inclination of the same bedding plane measured perpendicular to the strike.
Stromatolite	Sediment with a laminated structure formed by marine algae such as cyanobacteria. The resultant mounds are formed by alternating sediments and algal layers.
Stromatoporoid	A multi-celled colonial organism related to the sponges. Their calcareous skeletons contribute to reef building and are especially abundant in the Silurian limestones of Herefordshire.
Supercontinent	A major continental mass which consists of several joined blocks each of which were once separate continents in their own right (e.g. Pangaea).
Terrane	A fragment of crust, smaller than a plate, which now forms part of an orogenic belt but shows a distinct structural history supported by palaeo-magnetic data and/or fossil evidence of a derivation from some more distant location. These are sometimes referred to as exotic terranes, e.g. Anglesey in north Wales.

Till	The non-stratified material deposited directly by glacial ice which is poorly sorted and contains angular clasts which indicate little or no water transport. The term includes the material which forms the accumulations known as 'moraines'.
Topography	The general layout of a land surface including its size, relief and elevation.
Transcurrent fault	A fault, with a vertical or near vertical fault plane, where the principal displacement is horizontal rather than vertical. Synonymous with 'wrench fault' and 'strike-slip fault'.
Transform fault	A fault that laterally displaces mid-ocean ridges which at first appears to be a wrench fault (horizontal movement). However, movement in opposite directions only occurs between the offset segments of the ocean ridge, where the sides of the fault are parts of different oceanic plates. Beyond the sections of ocean ridge the fault continues but both sides are here part of the same plate and move in the same direction.

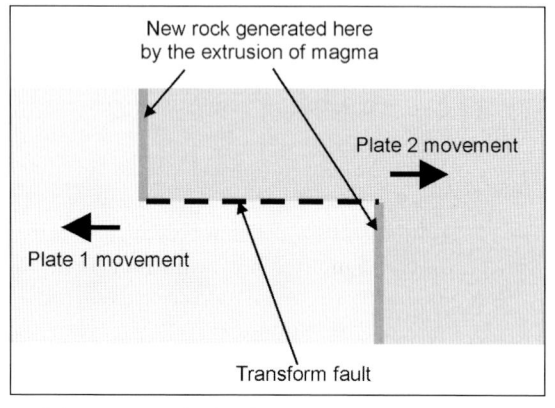

Transgression	An advance of the sea over the land (marine transgression) due to rapid rise in sea level caused by glacial melting or by subsidence of the land itself. Regression refers to the equivalent retreat of the sea.

References

Chapter 1
1. Christaller, W. *Die zentralen Orte in Suddeutschland* (Translated (in part), by Charlisle W. Baskin, as Central Places in Southern Germany. Prentice Hall 1966). This innovative work on the geography of settlements assumes a homogenous plain - something that distinguishes the Herefordshire Plain of Silurian/Devonian sediments - and the free movement of transport across it. The 'cornstone' hills of the central part of Herefordshire are scattered and do not significantly impede free movement across the plain: hence the distribution of major and minor settlements and the establishment of a 'hierarchy' in their relationships. (1933).
2. Olver, P.A. 'Old Red Sandstone: the Herefordshire stone' in *Transactions of the Woolhope Naturalists' Field Club*, **55**, (2007), pp.43-57.
3. Other fine viewpoints could include Bromyard Downs (in the north-east); the Black Mountains (in the west); Dinmore Hill (in the centre between Hereford and Leominster) and, when open, the Tower of Hereford Cathedral.

Chapter 2
1. H&W Earth Heritage Trust *Explore Geology and Landscape Trail Guide: Malvern Hills 1* (2003).
2. Masefield, J. '*Hereford Speech*' on being given the Freedom of the city of Hereford (1930).
3. Appleton, J. 'Some Thoughts on the Geology of the Picturesque' in J*ournal of Garden History,* **6** (3), (1986), pp.270-291.
4. Sinclair, J.B. and Fenn, R.W.D. 'Geology and the border squires' in *Transactions of the Radnorshire Society,* **69**, (1999), pp.143-172.
5. Murchison, R.I. '*The Silurian System, founded on geological researches in the counties of Salop, Hereford, …. Worcester and Stafford; with descriptions of the coal-fields and overlying formations*'. (London: John Murray), (1839) 768pp. (http://www2.odl.ox.ac.uk/gsdl/cgi-bin/library).
6. Turner, M.N. *Joseph Murray Ince (1806-1859): The Painter of Presteigne*, Logaston Press (2006).
7. As it happens, Farlow, on the north-east of Titterstone Clee Hill, was once a part of the county of Herefordshire – an outlying 'island' in Shropshire - so there is perhaps an historical reason for including this feature in a Geology of Herefordshire!
8. H&W Earth Heritage Trust *Explore Geology and Landscape Trail Guide::Wigmore Glacial Lake* (2001).
9. While some sources attribute the name Ambrey to the Romano-British leader, Ambrosius, it is interesting to note that as far back as 1896, members of the Woolhope Club had some uncertainties about this. At that time they began to think that the name may have been asso-

ciated with a landscape feature derived from the Celtic word 'Brea' or hill – as in Carn Brea in Cornwall. (Proceedings of the Woolhope Club: Anon 1898 pp. 124-25): cited in English Heritage Research Department Report Series No. 36-2008 ISSN 1749-8775.
10. Armstrong, P. *The English Parson-Naturalist: A Companionship Between Science and Religion,* Gracewing (2000).
11. Thackray, J.C. 'T.T. Lewis and Murchison's Silurian System' in *Transactions of the Woolhope Naturalists' Field Club,* **42** (2), (1977), pp.186-193.
12. Note the difference in spelling between the names of the village and the rock. This is probably due to a spelling mistake by Murchison.
13. Dreghorn, W. *Geology Explained in the Forest of Dean and the Wye Valley,* David & Charles (1968).
14. H&W Earth Heritage Trust *Explore Geology and Landscape Trail Guide: Wye Gorge* (2004).
15. H&W Earth Heritage Trust *Explore Geology and Landscape Trail Guide: Woolhope Dome* (2004).
16. H&W Earth Heritage Trust *Frome Valley Discovery Guide,* (2007), 23pp.

Chapter 4

1. Horner, L. 'On the mineralogy of the Malvern Hills' in *Transactions of the Geological Society of London* (Series 1), **1**, (1811), pp.281-321.
2. Brooks, M. 'Pre-Llandovery tectonism and the Malvern Structure' in *Proceedings of the Geologists' Association,* **81**, (1970), pp.249-268.
3. *ibid.*
4. Blyth, F.G.H. and Lambert, R.St.J. 'Chemical data from the Malvernian of the Malvern Hills, Herefordshire' in *Quarterly Journal of the Geological Society of London,* **125**, (1970), pp.543-555.
5. Brammall, A. 'Report of Easter Field Meeting at Hereford' in *Proceedings of the Geologists' Association,* **51**, (1940), pp.52-62.
6. Callaway, C. 'On the origin of the crystalline schists of the Malvern Hills' in *Quarterly Journal of the Geological Society of London,* **49**, (1893), pp.398-425.
7. Robertson, T. 'The section of the new railway tunnel through the Malvern Hills at Colwall' in *Summary of Progress of the Geological Survey of Great Britain and the Museum of Practical Geology* for the year 1925, (1926), pp.162-173.
8. Phipps, C.B. and Reeve, F.A.E. 'The Pre-Cambrian – Palaeozoic boundary of the Malverns' in *Geological Magazine,* **101**, (1964), pp.397-408.
9. Blyth, F.G.H. 'Malvern Tectonics – a contribution' in *Geological Magazine,* **89**, (1952), pp.185-194.
10. Phillips, J. 'The Malvern Hills compared with the Palaeozoic districts of Abberley, Woolhope, May Hill, Tortworth and Usk' *Memoir of the Geological Survey,* **2**, pt.1, (1848).
11. Lambert, R.St.J. and Holland, J.G. 'The petrography and chemistry of the Igneous Complex of the Malvern Hills, England' in *Proceedings of the Geologists' Association,* **82**, (1971), pp.323-352.
12. Watkins, Alfred *The Old Straight Track: Its Mounds, Beacons, Moats, Sites and Mark Stones,* Garnstone Press (1925).
13. Tucker, R.D. and Pharaoh, T.C. 'U-Pb zircon ages for Late Precambrian igneous rocks in southern Britain' in *Journal of the Geological Society of London,* **148**, (1991), pp.435-443.

14. Thorpe, R.S. 'Aspects of magmatism and plate tectonics in the Precambrian of England and Wales' in *Geological Journal*, **9**, (1974), pp.115-135.
15. Patchett, P.J., Gale, N.H., Goodwin, R. and Humm, M.J. 'Rb-Sr whole-rock isochron ages of late Precambrian to Cambrian igneous southern Britain' in *Journal of the Geological Society of London,* **137**, (1980), pp.649-656.
16. Woodcock, N.H. *The Precambrian and Silurian of the Old Radnor to Presteigne area,* in *Geological Excursions in Powys*, University of Wales Press (1993), pp.229-239.
17. Boynton, H.E. and Holland, C.H. 'Geology of the Pedwardine district, Herefordshire and Powys' in *Geological Journal*, **32**, (1997), pp.279-292.
18. Murchison, R.I. '*The Silurian System, founded on geological researches in the counties of Salop, Hereford, …. Worcester and Stafford; with descriptions of the coal-fields and overlying formations*'. (London: John Murray), (1839) 768pp. (http://www2.odl.ox.ac.uk/gsdl/cgi-bin/library).
19. Barclay W.J., Olver P., Hay S., Jenkins M., Payne M.J., Nicklin J. and Watkins N. 'The Precambrian inlier at Martley, Worcestershire: Martley Rock rediscovered' in *Transactions of the Woolhope Naturalists' Field Club*, **60**, (2012), pp.56-67.

Chapter 5

1. McKenzie, N.R., Hughes, N.C., Gill, B.C. and Myrow, P.M. 'Plate tectonic influences on Neoproterozoic - early Palaeozoic climate and animal evolution' in *Geology*, **42**, (2014), pp.127-130.
2. Earp, J.R. and Hains, B.A. *The Welsh Borderland*, British Regional Geology, British Geological Survey, HMSO, 3rd Edition (1971).
3. Holl, H.B. 'On the geological structure of the Malvern Hills and adjacent districts' in *Quarterly Journal of the Geological Society of London*, **21**, (1865), pp.72-102.
4. Oliver, P.G and Payne, M.J.P. 'Improved exposures of the Precambrian/Lower Palaeozoic contacts in the Malvern Hills' in *Proceedings of the Cotteswold Naturalists' Field Club*, **43**, (2004), pp.84-91.
5. Worssam, B.C., Ellison, R.A. and Moorlock, B.S.P. *Geology of the country around Tewkesbury*, Memoir of the British Geological Survey, sheet 216 (England & Wales). HMSO, London (1989).
6. Blyth, F.G.H. 'The basic intrusive rocks associated with the Cambrian inlier near Malvern' in *Quarterly Journal of the Geological Society of London*, **91**, (1935), pp.463-478.
7. Boynton, H.E. and Holland, C.H. 'Geology of the Pedwardine district, Herefordshire and Powys' in *Geological Journal*, **32**, (1997), pp.279-292.
8. Toghill, P. 'The Shelveian Event, a late Ordovician tectonic episode in southern Britain (eastern Avalonia)' in *Proceedings of the Geologists' Association*, **103**, (1992), pp.31-35.
9. Brooks, M. 'Pre-Llandovery tectonism and the Malvern Structure' in *Proceedings of the Geologists' Association*, **81**, (1970), pp.249-268.

Chapter 6

1. Murchison, R.I. '*The Silurian System, founded on geological researches in the counties of Salop, Hereford, …. Worcester and Stafford; with descriptions of the coal-fields and overlying formations*'. (London: John Murray), (1839) 768pp. (http://www2.odl.ox.ac.uk/gsdl/cgi-bin/library).
2. Lawson, J.D., Curtis, M.L.K., Squirrell, H.C., Tucker, E.V. and Walmsley, V.G. 'The Silurian inliers of the south-eastern Welsh Borderland' in *Geologists' Association Guide*, no. 5, (1982), 33pp.

3. Squirrell, H.C. and Tucker, E.V. 'The geology of the Woolhope Inlier, Herefordshire' in *Quarterly Journal of the Geological Society of London*, **116**, (1960), pp.139-185.
4. Lawson, J.D. 'The Silurian succession at Gorsley (Herefordshire)' in *Geological Magazine*, **91**, (1954), pp.227-237.
5. Lawson, J.D. 'The geology of the May Hill area' in *Quarterly Journal of the Geological Society of London*, **111**, (1955), pp.85-116.
6. Siveter, D.J., Owens, R.M. and Thomas, A.T. *Silurian Field Excursions: A geotraverse across Wales and the Welsh Borderland*, National Museum of Wales Geology Series No.**10** Cardiff (1989).
7. Phipps, C.B. and Reeve, F.A.E. 'Stratigraphy and geological history of the Malvern, Abberley and Ledbury Hills' in *Geological Journal*, **5**, (1967), pp.1-37.
8. Ziegler, A.M., Cocks, L.R.M. and McKerrow, W.S. 'The Llandovery transgression of the Welsh Borderland' in *Palaeontology*, **11**, (1968), pp.736-782.
9. e.g. Holland, C.H., Lawson, J.D. and Walmsley, V.G. 'The Silurian rocks of the Ludlow district, Shropshire' in *Bulletin of the British Museum (Natural History) Geology*, **8,** No. 3 (1963);
Holland, C.H. and Bassett, M.G. (eds.) 'A global standard for the Silurian System, National Museum of Wales', *Geology Series* No. 9, Cardiff (1989).
10. Rosenbaum, M. 'The future for geology in the Marches' in *Proceedings of the Shropshire Geological Society*, **13**, (2008), pp.100-103.
11. Siveter, D.J. 'Pitch coppice' in Aldridge, R.J., Siveter, D.J., Siveter, D.J., Lane, P.D., Palmer, D. and Woodcock, N.H., British Silurian stratigraphy, *Geological Conservation Review Series* **19**, (2000), Chapt. 5 (http://www.thegcr.org.uk/ImageBank.cfm?v=19&Style=Chapter&Chapter=05).
12. Phillips, J. 'On the occurrence of shells and corals in a conglomerate bed, adherent to the trap rocks of the Malvern Hills, and full of rounded and angular fragments of those rocks.' in *Philosophical Magazine*, **21**, (1842), pp.288-293.
13. Siveter, D.J. 'The Silurian Herefordshire Konservat-Lagerstätte: a unique window on the evolution of life' in *Proceedings of the Shropshire Geological Society*, **13**, (2008), pp.58-61.
14. Siveter, D.J., Sutton, M.D., Briggs, D.E.G. and Siveter, D.J. 'A Silurian sea spider' in *Nature* **431**, (2004), pp.978-980.
15. Siveter, D.J., Briggs, D.E.G., Siveter, D.J. and Sutton, M.D. 'A 425–million-year-old Silurian pentastomid parasitic on ostracods' in *Current Biology*, **25**, (2015), pp.1-6. (http://dx.doi.org/10.1016/j.cub.2015.04.035)
16. Whitaker, J.H.McD. 'The geology of the area around Leintwardine' in *Quarterly Journal of the Geological Society of London*, **114**, (1962), pp.319-351.
17. Alexander, F.E.S. 'The Aymestry Limestone of the main outcrop' in *Quarterly Journal of the Geological Society of London*, **92**, (1936), pp.103-115.
18. Brodie, P.B. 'On the occurrence of remains of Eurypterus and Pterygotus in the Upper Silurian rocks in Herefordshire' in *Quarterly Journal of the Geological Society of London*, **25**, (1869), pp.235-237.

Chapter 7
1. Hearne, T. *A scene from a window at Moccas Court*. (c.1788-9) http://www.britishmuseum.org/research/collection_online/collection_object_details/collection_image_gallery.aspx?assetId=184110001&objectId=749020&partId=1.

2. Clarke, B.B. 'A note on the geology of the Deepwell tufa at Moccas and Brobury Scar' in *Transactions of the Woolhope Naturalists' Field Club*, **33**, (1952), p.48.
3. Sheridan, R.E. 'Pulsation tectonics as a control of long-term stratigraphic cycles,' in *Palaeoceanography*, 2, (1987), pp.97-118.
4. Barclay, W.J., Brown, M.A.E., McMillan, A.A., Pickett, E.A., Stone, P., & Wilby, P.R. *The Old Red Sandstone of Great Britain*, Geological Conservation Review Series No.31, Joint Nature Conservation Committee, (2005), p.15.
5. Allen, J.R.L. *Pedogenic calcretes in the Old Red Sandstone facies (late Silurian-early Carboniferous) of the Anglo-Welsh area, southern Britain* in Wright, V.P. (ed.) Palaeosols: their recognition and interpretation, (1986), pp.56-86.
6. Richardson, J.B., Rodriguez, R.M. & Sutherland, S.J.E. 'Palynology and recognition of the Silurian/Devonian boundary in some British terrestrial sediments by correlation with Cantabrian and other European marine sequences - a progress report' in *Courier Forschungsinstitut Senckenberg*, 220, (2000), pp.1-7.
7. Barclay, W.J., Davies, J.R., Hillier, R.D. & Waters, R.A. *Lithostratigraphy of the Old Red Sandstone successions of the Anglo-Welsh Basin*, British Geological Survey Research Report, RR/14/02 (2015).
8. Richardson *et al* 'Palynology', *op. cit.*
9. Barclay *et al Lithostratigraphy*, *op. cit.*
10. Brandon, A. G*eology of the country between Hereford and Leominster*. Memoir of the British Geological Survey, Sheet 198 England and Wales (1989), 62 pp.
11. Allen, J.R.L. *Pedogenic calcretes in the Old Red Sandstone*, *op. cit.*, note 5 above.
12. Parker, A., Allen, J.R.L. & Williams, B.P.J. 'Clay mineral assemblages of the Townsend Tuff Bed (Lower Old Red Sandstone), South Wales and the Welsh Borders' in *Journal of the Geological Society of London*, **140,** (1989), pp.769-779.
13. Clarke, B.B. 'The Old Red Sandstone of the Merbach Ridge, Herefordshire, with an account of the Middlewood Sandstone, a new fossiliferous horizon 500 feet below the Psammosteus Limestone' in *Transactions of the Woolhope Naturalists' Field Club*, **34**, (1954), pp.195-219.
14. Allen, J.R.L. & Williams, B.P.J. 'Beaconites antarcticus: a giant channel-associated trace fossil from the Lower Old Red Sandstone of South Wales and the Welsh Borders' in *Geological Journal*, **16**, (1981), pp.255-269.
15. Allen, J.R.L. & Crowley, S.F. 'Lower Old Red Sandstone fluvial dispersal systems in the British Isles' in *Transactions of the Royal Society of Edinburgh, Earth Sciences*, **74**, (1983), pp.61-68.
16. Barclay, W.J., Rathbone, P.A., White, D.E. & Richardson, J.B. 'Brackish water faunas from the St. Maughans Formation: the Old Red Sandstone section at Ammons Hill, Hereford and Worcester, UK, re-examined' in *Geological Journal*, **29**, (1994), pp.369-379.
17. Bluck, P.J., Cope, J.C.W. & Scrutton, C.T. 'Devonian' in Cope, C.J., Ingham, J.K. & Rawson, P.F. (eds.) in *Atlas of Palaeogeography and Lithofacies*, Geological Society, London, Memoir (1992) 13, pp.57-66.
18. Dineley, D. and Metcalf, S. *Devonian Fossil Fishes sites of the Welsh Borders*, Chapter 4 in *Fossil Fishes of Great Britain*, Geological Conservation Review Series, Volume 16 (Joint Nature Conservation Committee) (1999).
19. Allen, J.R.L. 'Studies in fluviatile sedimentation: Bars, bar-complexes and sandstone sheets (low-sinuosity braided streams) in the Brownstones (L Devonian), Welsh Borders' in *Sedimentary Geology*, **33** (2), (1983), pp.237-293.

20. Allen, J.R.L. 'Source rocks of the Lower Old Red Sandstone: Exotic pebbles from the Brownstones, Ross-on-Wye, Hereford and Worcester' in *Proceedings of the Geologists' Association*, **85**, (1974), pp.493-510.
21. Love, S.E. & Williams, B.P.J. *Sedimentology, cyclicity and floodplain architecture in the Lower Old Red Sandstone of SW Wales* in Friend, R.F. and Williams, B.P.J. (eds) in New Perspectives on the Old Red Sandstone. Geological Society, London, Special Publications, **180**, (2000), pp.371-388.
22. Thurlby, M. 1999 *The Herefordshire School of Romanesque Sculpture* Logaston Press (1999), new edition 2013.
23. Raup, D.M. *Extinction - Bad Genes or Bad Luck?* Norton (1991). (General overview of mass extinction theories).
24. Sepkoski, J.J. 'Mass extinctions in the Phanerozoic oceans: a review' in Silver, L.T. and Schultz, P.H. (eds.) *Geological Implications of Impacts of Large Asteroids and Comets on the Earth*, Geological Society of America Special Paper, **190**, (1982), pp.283-289.
25. House, M.R. 'Correlation of Mid-Palaeozoic ammonoid evolutionary events with global sedimentary perturbations' in *Nature*, **313**, (1985), pp.17-22.
26. Rondot, J. 'La structure de Charlevoix comparée à d'autres impacts météoritiques' in *Canadian Journal of Earth Sciences*, **7**, (1970), pp.1194-1202.
27. McGhee, G.R. Jnr. *The Late Devonian mass extinction: The Famennian/Frasnian crisis* in Critical Moments in Palaeobiology and Earth History Series, Columbia University Press, New York (1996), 303pp.
28. Dineley, D.L. *Aspects of a Stratigraphic System: The Devonian*, Macmillan (1984), 223pp.

Chapter 8

1. Mushet, D. 1824 'Vertical section of the strata of the Forest of Dean, constructed from actual sinkings and from observations made at the surface', in Buckland W. and Conybeare, W.D. 'Observations on the south western coal district of England', in *Transactions of the Geological Society*, (2nd Series), **1**, (1824), pp.210-316
2. Trotter F.M. *Geology of the Forest of Dean coal and iron ore field*. Memoir of the Geological Survey of Great Britain, Sheet 233 (1942)
3. Waters, C.N., Browne, M.A.E., Dean, M.T., and Powell, J.H. *Lithostratigraphical framework for Carboniferous successions of Great Britain (Onshore)*. British Geological Survey Research Report, RR/09/01 (2009).
4. *ibid.*
5. Clark, N.D.L., Gillespie, R., Morris, S.F. and Clayton, G. 'A new early Carboniferous crustacean from the Forest of Dean, England' in *Journal of Systematic Palaeontology*, **14**, (2016), pp.799-807.
6. Barclay, W.J. and Smith, N.J.P. *Geology of the country between Hereford and Ross-on-Wye (Sheet 215 Ross-on-Wye)*, British Geological Survey; Sheet Explanation for Sheet 215, (2002), 29pp.
7. Worssam, B.C., Ellison, R.A., Moorlock, B.S.P. *Geology of the Country around Tewkesbury*. Memoir of the British Geological Survey, Sheet 216 (1989).
8. Smith D.B. and Taylor, C.M. 1992 *Permian* in Cope, J.C.W., Ingham, J.K. and Rawson, P.F. (eds) *Atlas of Palaeogeography and Lithofacies*, Memoir of the Geological Society of London, No 13, (1992).
9. Barclay, W.J., Ambrose, K., Chadwick, R.A. and Pharaoh, T.C. *Geology of the country around Worcester*. Memoir of the British Geological Survey, Sheet 199 (p101), (1997).

10. *ibid.*
11. *ibid.*
12. Blackith, R.E. 'The Haffield Breccias' in *Scientific Journal of the Royal College of Science*, **26**, (1956), pp.77-85.
13. Hounslow, M.W., McKie, T. and Ruffell, A.H. 'Permian to Late Triassic post-Orogenic Collapse and Rifting' in Woodcock N.H. and Strachan, R.A. (eds) *Geological History of Britain and Ireland*, Wiley-Blackwell (2012).
14. Ambrose, K., Hough, E., Smith, N.J.P. and Warrington, G. *Lithostratigraphy of the Sherwood Sandstone Group of England, Wales and south-west Scotland*, British Geological Survey Research Report, RR/14/01 (2014).

Chapter 9

1. Moreau, M., Shand, P., Wilton, N., Brown, S. and Allen, D. *Baseline Report Series 12: The Devonian Sandstone aquifer of south Wales and Herefordshire.* British Geological Survey Commissioned Report No. CR/04/185N (2004).
2. Phillips, J. *The Malvern Hills compared with the Palaeozoic districts of Abberley, Woolhope, May Hill, Tortworth and Usk*, Memoir of the Geological Survey of GB, **2**, (1), (1848), p.14.
3. Brooks, M. 'Some aspects of the Paleogene evolution of western Britain in the context of an underlying mantle hot spot' in *Journal of Geology*, **81** (1), (1973), pp.81-88.
   Cope, J.C.W. 'Cretaceous Hotspot and Tilt of Britain' in *Journal of the Geological Society, London*, **151**, (1994), pp.905-908.
4. Gibbard, P.L. and Lewin, J. 'The history of the major rivers of southern Britain during the Tertiary' in *Journal of the Geological Society of London*, **160**, (2003), pp.829-845.
5. Brown, E.H. *The Relief and Drainage of Wales*, University of Wales Press (1960).
   Richards, A.E. 'Glaciation and drainage evolution in the southern Welsh Borderland' in *Proceedings of the Shropshire Geological Society*, **13**, (2008), pp.92-99.
6. Sparks, B.W. *Geomorphology*, Longmans (1965).
   H&W Earth Heritage Trust *Explore Geology and Landscape Trail Guide: Wye Gorge* (2004).
   Dreghorn, W. *Geology Explained in the Forest of Dean and the Wye Valley*, David & Charles (1968).
7. Sparks, B.W. (1965) *ibid.*
   Dreghorn, W. (1968) *ibid.*
   H&W Earth HeritageTrust Explore Geology and Landscape Trail Guide: *Symonds Yat* (2001).
   Miller, A. 'Entrenched Meanders of the Herefordshire Wye' in *Geographical Journal*, **85**, 2, (1935), pp158-178.
8. H&W Earth Heritage Trust *Explore Geology and Landscape Trail Guide: Wye Gorge* (2004).
9. Apsimon, A. 'Getting it right: No Middle Palaeolithic at King Arthur's Cave' in *Proceedings of the University of Bristol Spelaeological Society*, **23**, (1), (2003), pp.17-26.
   Barton, R.N.E., Price, C. and Proctor, C. 'The Wye Valley Caves project: recent investigations into King Arthur's Cave and the Madawg rockshelter' in Lewis, S.G. and Maddy, D. *Quaternary of the south Midlands and Welsh Marches: Field guide*, London: Quaternary Research Association (1997), pp.63-75.
10. Barclay, W.J. and Wilby, P.R. *Geology of the Talgarth District. A brief explanation of the geological map Sheet 214 Talgarth*, British Geological Survey (2003).
11. Richards, A.E. 'Re-evaluation of the Middle Pleistocene Stratigraphy of Herefordshire' in *Journal of Quaternary Science*, **13**, (1998), pp.115-136.

Richards, A.E. 'Middle Pleistocene glaciation in Herefordshire: the sedimentology and structural geology of the Risbury Formation (Older Drift Group)' in *Proceedings of the Geologists' Association*, **110**, (1999), pp.173-192.

Richards, A.E. 'Herefordshire' in Lewis C.A. and Richards, A.E. (eds) *The glaciations of Wales and adjacent areas*, Logaston Press (2005).

12. Coope, G.R., Field, M.H., Gibbard, P.L., Greenwood, M. and Richards, A.E. 'Palaeontology and biostratigraphy of Middle Pleistocene river sediment in the Mathon Member, at Mathon, Herefordshire' in *Proceedings of the Geologists' Association*, **111** (3), (2002), pp.237-258.

13. H&W Earth Heritage Trust *Explore Geology and Landscape Trail Guide: Symonds Yat* (2001).

    H&W Earth Heritage Trust *Explore Geology and Landscape Trail Guide: Wye Gorge* (2004).

14. Barclay, W.J., Brandon, A., Ellison, R.A. and Moorlock, B.S.P. 'A Middle Pleistocene palaeovalley fill west of the Malvern Hills' in *Journal of the Geological Society, London*, **149**, (1992), pp.75-92.

    Brandon, A. *Geology of the country between Hereford and Leominster*, Memoir of the Geological Survey of Great Britain, Sheet 198, (1989).

    Richards, A.E. 'The Pleistocene stratigraphy of Herefordshire', unpublished PhD thesis, University of Cambridge (1994), and note 11 above.

15. Barclay *et al.* (1992) *ibid.*

16. H&W Earth Heritage Trust *Explore Geology and Landscape Trail Guide: Kington and Hergest* (2007).

    Luckman, B.H. 'The Hereford Basin' in Lewis, C.A. (ed), *The glaciations of Wales and adjoining regions*, Longman (1970), pp.175-196.

17. Richards, A.E. 'Middle Pleistocene glaciation in Herefordshire' in Lewis, S.G. and Maddy, D. (Editors) *Quaternary of the South Midlands and the Welsh Marches: Field Guide*, Quaternary Research Association, Cambridge (1997) and note 11 above.

18. Barclay, W.J., Brandon, A., Ellison, R.A. and Moorlock, B.S.P. 'A Middle Pleistocene palaeovalley fill west of the Malvern Hills' in *Journal of the Geological Society, London*, **149**, (1992), pp.75-92.

    Richards, A.E. 'Sedimentological and geomorphological evidence for the development of Herefordshire's river system' in *Transactions of the Woolhope Naturalists' Field Club*, **55**, (2007), pp.61-78.

    Luckman, B.H. 'The Hereford Basin' in Lewis, C.A. (ed), *The glaciations of Wales and adjoining regions*, Longman (1970), pp.175-196.

19. Brandon, A. *Geology of the country between Hereford and Leominster*, Memoir of the Geological Survey of Great Britain, Sheet 198 (1989).

20. Symonds, W.S. *Records of the Rocks: Notes on the Geology, Natural History and Antiquities of North and South Wales, Devon and Cornwall* John Murray (1872).

    Richardson, L. 'An outline of the geology of Herefordshire' in *Transactions of the Woolhope Naturalists' Field Club*, Vol. for 1905-1907, (1911), pp.1-68.

21. Cross, P. 'Glacial and proglacial deposits, landforms and river diversions in the Teme valley near Ludlow, Shropshire', PhD Thesis, Univ. London (1971), 407pp.

    Cross, P. *The glacial geomorphology of the Wigmore and Presteigne basins and some adjacentarea*, MSc Thesis, Univ. London, (1966) 86pp.

    Dorling, P. *The Lugg Valley, Herefordshire: Archaeology, Landscape Change and Conservation* (2007).

Cross, P. and Hodgson, J.M. 'New evidence for the glacial diversion of the River Teme near Ludlow, Salop' in *Proceedings of the Geologists' Association*, **3**, (1975), pp.313-331.
22. Toghill, P. *Geology of Shropshire*, 2nd edition, The Crowood Press (2006), p.234.
23. Richards, A.E. 'Middle Pleistocene glaciation in Herefordshire' in Lewis, S.G. and Maddy, D. (Editors) *Quaternary of the South Midlands and the Welsh Marches: Field Guide*, Quaternary Research Association, Cambridge (1997).
Richards, A.E. 'Herefordshire' in Lewis C.A. and Richards, A.E. (eds) *The glaciations of Wales and adjacent areas*, Logaston Press (2005).
24. Dwerryhouse, A.R. and Miller, A.A. 'Glaciation of Clun Forest, Radnor Forest and some adjoining districts' in *Quarterly Journal of the Geological Society of London*, **86**, (1930), pp.96-129.
25. Stokes, K. 'The Late Quaternary vegetation history of the southern Welsh borderland' Unpublished M Phil. Thesis, Kingston University (2003).
26. Luckman, B.H. 'Some aspects of the geomorphology of the Lugg and Arrow valleys', unpublished MA thesis, Manchester University (1966).
27. Brandon, A. *Geology of the country between Hereford and Leominster*, Memoir of the Geological Survey of Great Britain, Sheet 198, (1989).
28. Cross, P. 'Some aspects of the glacial geomorphology of the Wigmore and Presteigne districts' in *Transactions of the Woolhope Naturalists' Field Club*, **39** (2), (1968), pp.198-221.
H&W Earth Heritage Trust *Explore Geology and Landscape Trail Guide: Wigmore Glacial Lake* (2001).
29. Rosenbaum, M.S. 'A Geological Trail in front of the last glacier in South Shropshire.' in *Proceedings of the Shropshire Geological Society*, **12**, (2007), pp.56-69.
30. Richards, A.E. 'Middle Pleistocene glaciation in Herefordshire' in Lewis, S.G. and Maddy, D. (Editors) *Quaternary of the South Midlands and the Welsh Marches: Field Guide*, Quaternary Research Association, Cambridge (1997).
Cross, P. and Hodgson, J.M. 'New evidence for the glacial diversion of the River Teme near Ludlow, Salop' in *Proceedings of the Geologists' Association*, **3**, (1975), pp.313-331.
31. Dwerryhouse, A.R. and Miller, A.A. 'Glaciation of Clun Forest, Radnor Forest and some adjoining districts' in *Quarterly Journal of the Geological Society of London*, **86**, (1930), pp.96-129.
32. Barclay, W.J. and Wilby, P.R. *Geology of the Talgarth District, a brief explanation of the geological map Sheet 214 Talgarth*. British Geological Survey (2003).
33. Bryant, R. 'Rock slope failure and glaciation of the Darens, Olchon Valley, Herefordshire' in *Transactions of the Woolhope Naturalists' Field Club*, **58**, (2010), pp.127-140.
Hilton, K. *Process and Pattern in Physical Geography*, University Tutorial Press (1987), pp.70, 72, 91-93, 105-107, 149-152.
34. Brandon, A. *Geology of the country between Hereford and Leominster*, Memoir of the Geological Survey of Great Britain, Sheet 198, (1989).
H&W Earth Heritage Trust *Frome Valley Geology and Landscape Discovery Guide* (2007).
35. Hopkinson, C. and Jenkins, M. 'A Landslip at Shortwood' in *Earth Matters (Newsletter of the Woolhope Club Geology Section)*, no. 11, p.6. (2014).
36. Dury, G.H. *Subsurface exploration and chronology of underfit streams*. Geological Survey Professional Paper no. 452-B, (1964), pp.236-295.
Dury, G.H. *The Face of the Earth*, Pelican Books, (1969).

37. Dury, G.H. *Theoretical implications of underfit streams*. Geological Survey Professional Paper no 452-C. US Govt. Printing Office, Washington DC, (1965).
38. H&W Earth Heritage Trust *Explore Geology and Landscape Trail Guide: Byton and Kinsham* (2004).
39. Dreghorn, W. *Geology Explained in the Forest of Dean and the Wye Valley*, David & Charles (1968); Hilton, K. *Process and Pattern in Physical Geography*, UniversityTutorial Press (1987), pp.70, 72, 91-93, 105-107, 149-152.
   H&W Earth Heritage Trust *Explore Geology and Landscape trail guide*: Ross-on-Wye (2004)
40. Hilton, K. *ibid.*
41. H&W Earth Heritage Trust *Explore Geology and Landscape Trail Guide: Woolhope Dome* (2004).
42. H&W Earth Heritage Trust *Frome Valley Geology and Landscape Discovery Guide* (2007).
43. Brandon, A. *Geology of the country between Hereford and Leominster*, Memoir of the Geological Survey of Great Britain, Sheet 198, (1989).
44. H&W Earth Heritage Trust *Explore Geology and Landscape Trail Guide: Wye Gorge* (2004). Harding, B. 'Tufa formation today and in the past' in *Transactions of the Woolhope Naturalists' Field Club*, **49** (2), (1998), pp.70-181.
45. H&W Earth Heritage Trust *Explore Geology and Landscape Trail Guide: Queenswood & Bodenham* (2004).
46. Gregory, K.J. *Fluvial geomorphology of Great Britain*, Geological Conservation Review, **13**, (1997), pp.299-304 (River Lugg).
47. H&W Earth Heritage Trust *Explore Geology and Landscape Trail Guide: Hampton Bishop* (2004).

Chapter 10

1. Wall, T., Weightman, J. and Davey, S. *Downton Gorge National Nature Reserve*, Woolhope Naturalists' Field Club (2011), p.7.
2. Torrens, H. *Reprint of, and Additional Material: J. Phillips. (1844) Memoirs of William Smith LLD*, The Bath Royal Literary and Scientific Institution (2003), p.230.
3. Cantrill, T.C. 'On a boring for coal at Presteign, Radnorshire' in *Geological Magazine*, **4**, (1917), pp.481-492.
   Cantrill, T.C. 'Boring for coal at Presteign (Corr)' in *Geological Magazine*, **5**, (1918), pp.47-48.
   Watts, W.W. 'Coal in the Silurian at Presteign (Corr)' in *Geological Magazine*, **4**, (1917), pp.552-553.
4. Leake, R.C. *et al*, *The potential for gold mineralisation in the British Permian and Triassic red beds and their contacts with underlying rocks*, Report of the British Geological Survey, Mineral Reconnaissance (144), (1997).
5. Hart, C.E. Gold in Dean Forest, *Transactions of the Bristol and Gloucester Archaeological Society*, **65**, (1944), pp.98-104.
6. Murchison, R.I. *The Silurian System, founded on geological researches in the counties of Salop, Hereford, …. Worcester and Stafford; with descriptions of the coal-fields and overlying formations*, John Murray (1839), p.204 (http://www2.odl.ox.ac.uk/gsdl/cgi-bin/library).
7. Watkins, A. 'Herefordshire Pipe Factories: Pipe Aston' in *Transactions of the Woolhope Naturalists' Field Club*, (1931), pp.132-3.
8. Goodbury, V. *Herefordshire Limekilns*, University of Birmingham Dissertation (Copy in Hereford Library) (1992).

9. Brooks, A. *Herefordshire*, Pevsner Architectural Guides: Buildings of England, Yale University Press (2012).
   Bloodworth, A.J., Cameron, D.G., Harrison, D.J., Highley, D.E., Holloway, S. and Warrington, G. Mineral Resource Information for Development Plans – Phase one Herefordshire & Worcestershire: resources and constraints, Report of the British Geological Survey, Mineral Resources Series, no. WF/99/04, (1999), 37pp.
   English Heritage *A building stone atlas of Herefordshire*, (2012), 21pp.
10. Herefordshire & Worcestershire Earth Heritage Trust Leinthall Earls Quarry [Online] http://www.earthheritagetrust.org/pub/learning-discovery/aggregates/aggregates-of-hereford-shire/site-examples-hfds/leinthall-earls-quarry/ (Accessed 2016).
11. Barclay, W.J., Ambrose, K., Chadwick, R.A. and Pharaoh, T.C. *Geology of the country around Worcester*, Memoir of the Geological Survey of Great Britain, Sheet 199, (1997), p.122.
12. Coplestone-Crow, B. *Herefordshire Place-Names*, Logaston Press (2009), pp.64, 200.
13. Eisel, J. 'Aspects of the Wye navigation' in *Transactions of the Woolhope Naturalists' Field Club*, **60**, 2012, pp.47-53.
14. Daniels, Stephen and Watkins, Charles (eds) *The Picturesque Landscape: Visions of Georgian Herefordshire* Dept. of Geography, University of Nottingham in association with Hereford City Art Gallery and University Art Gallery, Nottingham (1994).
15. see *Marches* by Andrew Allott, New Naturalist series No. 118, Harper-Collins (2011).
16. see *Wye Valley* by George Peterken, New Naturalist series No. 105, Harper-Collins (2008).
17. Appleton, Jay 'Some thoughts on the geology of the picturesque' in *Journal of Garden History*, **6**, no.3, (1986), pp.270-291.
18. see *The Secrets of Downton Gorge* – a 60 min. DVD with excellent sections on the geology, natural history and conservation in the Gorge produced by Libraprim SA, Berengere Primat-Serval, 210 route de Jussy, CH-1243 Presinge, Geneva, Switzerland (original video 1993).
19. Met Office (Meteorological Office) *UK climate anomaly maps* (Online) http://www.metoffice.gov.uk/climate/uk/anomacts/ (Accessed 2015).
20. Mackney, D. and Burnham, C.P. 'The soils of the West Midlands' in *Bulletin of the Soil Survey of Great Britain*, (2), (1964), 111pp.
21. Beard, G.R. *Soils of Worcester and the Malverns district*, Memoir of the Soil Survey of Great Britain, (1986), 105pp.
22. HWT (Herefordshire Wildlife Trust) *Wildlife Reserves* (Online) http://www.herefordshirewt.org/wildlife/reserves/ (Accessed 2016).
23. HTT (Herefordshire through Time, Herefordshire County Council) (Online) *Trench Royal, Vowchurch* (SMR 365) http://htt.herefordshire.gov.uk/her-search/monuments- search/search/10monument?smr_no=365&s=Start+search (Accessed 2016).
24. Richardson, L. *Wells and springs of Herefordshire*, Memoir of the Geological Survey of Great Britain, (1935), 136pp.
25. Harvey, T. and Gray, J. The unconventional hydrocarbon resources of Britain's onshore basins - Shale gas, Department of Energy and Climate Change report (2012), 35pp.
26. Rollin, K.E. 'Low-temperature geothermal energy' in *Energy resources*, British Geological Survey, (19), (2003), pp.24-25.
    Barker, J.A. *et al*, 'Hydrogeothermal studies in the United Kingdom' in *Quarterly Journal of Engineering Geology and Hydrogeology*, **33**, (2000) pp.41-58.
27. Murchison, R.I. 'The Silurian System, founded on geological researches in the counties of Salop, Hereford, …. Worcester and Stafford; with descriptions of the coal-fields and overlying formations'.

(London: John Murray), (1839) 768pp. (http://www2.odl.ox.ac.uk/gsdl/cgi-bin/library), p.435.

Symonds, W.S. '"The Wonder", near Marcle' in *Transactions of the Woolhope Naturalists' Field Club*, (1878), pp.74-5.

28. Brandon, A. *Geology of the country between Hereford and Leominster*, Memoir of the Geological Survey of Great Britain, Sheet 198, (1989), p.44.

29. ibid. p.25.

    Moore, H.C., Clarke, R. and Watkins, A. 'The Earthquake of December 17th, 1896' in *Transactions of the Woolhope Naturalists' Field Club* (1896), pp.228-235.

    Davison, C. *The Hereford Earthquake of December 17, 1896*, Cornish Bros. (1899), 303pp.

30. Rees, D.M., Bradley, E.J. and Green, B.M.R. *Radon in Homes in England and Wales: 2010 Data Review*, Report no. HPA-CRCE-015, Centre for Radiation, Chemical and Environmental Hazards, (2011), p.153. (Kington Post Code HR5).

31. Zalasiewicz, J. and Williams, M. *The Goldilocks Planet*, Oxford University Press (2012), pp.162-3.

INDEX

Acadian orogeny, 128, 139, 146, 211
agriculture, 203-5
Anglian, 173, 174, 178, 201
Ankerdine Hill, 97
archaeology, 195-6
    hand axe, 195
    hillfort, 3, 7, 12, 18, 21, 28, 71, 73, 74, 108, 202
    King Arthur's Cave, 21, 26, 148, 171
    King Arthur's Stone, 202
    rock shelter, 201
    sites, 21
Artistic movements, 12, 17, 24, 27, 32, 123, 142, 203
Avalonia, 43-44, 57, 59, 93, 94
Avon Group, 147, 148, 152, 155
Aymestrey, 17, 20, 22, 31, 32, 116
Aymestry Limestone, 19, 23, 27, 28, 31, 100, 107, 108, 116, 117, 183, 186

Backbury Hill, 28, 31
Banks family, *ix-x*, 12, 16, 17
Barry Harbour Limestone, 148, 149, 193
Bartestree, 30, 31, 156
bentonite, 28, 47, 103, 110, 186, 197
Biblins, 26, 137, 170, 175, 176, 193
bioturbation, 115, 118, 119
Bircher Common, 12, 19
Bishop's Frome, 154, 192
Bishop's Frome Limestone. See Chapel Point Limestone
Black Darren, 124, 185
Black Hill. See Cat's Back
Black Mountains, 3, 10, 11, 12, 13, 15, 18, 24, 31, 123, 124, 172, 185, 203, 204
    Drainage, 172
    Rocks, 33, 136, 138, 165, 166
    Topography, 13, 124, 172
Bodenham, 144, 175, 193, 201
brachiopod, 23, 81, 83, 86, 87, 92, 97, 102, 103, 105, 106, 107, 110, 112, 113, 115, 118, 119, 120, 121, 134, 147
    Kirkidium, 23, 92
    Lingula, 97
Bradnor Hill, 11-17, 19, 78, 97, 113, 114, 118, 120, 166, 176, 180
    Quarries, 198

Brampton Bryan, 77
Bredwardine, 15, 21, 193, 198
Breinton, 123, 131, 183
Bridgnorth Sandstone, 160, 161, 162, 163
Bringewood Forge, 120, 196
British Camp. See Herefordshire Beacon
British Geological Survey, 100, 101, 103, 121
Brobury Scar, *xiv*, 123
Bromsgrove Sandstone Formation. See Helsby Sandstone Formation
Bromyard, 2, 197
    Downs, 12, 132, 134
    Hackley Farm, 133
    Linton Tile Works, 133
    Plateau, 165, 174, 175, 176, 192
Bronsil Castle, 101
Bronsil Shale, 87, 89
Brownstones Formation, 137, 138, 139, 154, 165, 172, 175, 200
bryozoan, 92, 107, 109, 112
building stone, 5, 56, 79, 115, 116, 120, 130, 138, 142, 159, 160, 163, 193, 198, 199, 200
Byton, 185, 187, 189

Cadomian orogeny, 57, 63, 74
Cainozoic Era, 165-94
calcrete, 126, 127, 130, 131, 132, 133, 136, 138, 167, 185, 186, 204
Caledonian orogeny, 44, 59, 124, 128, 139, 212
Cambrian Period, 35, 43, 53, 58, 59, 81-87, 94, 100, 101, 110, 124
Carboniferous limestone, 21, 24, 25, 45, 137, 150, 151, 152, 155, 165, 166, 170, 176, 198, 201, 203, See also Pembroke Limestone Group
Carboniferous Period, 24, 35, 45, 59, 60, 79, 88, 97, 145-57
Cat's Back, 14, 124, 136, 137, 204
Chances Pitch, 116
Chapel Point Limestone, 126, 127, 130, 131, 135, 136, 167, 193, 198
Chase End Hill, 10, 75, 90
Chase Wood, 155, 190, 191
Checkley, 106
chronology. See time, geological
Church Stretton Fault, 15, 17, 58, 77-79, 81-83, 88, 89, 95, 97, 102, 107, 113, 119, 153, 159, 165

229

Clifford's Mesne, 102
Clutter's Cave, 76, 77
coal, 46, 54, 145, 152, 196, 197
Coalbrookdale Formation, 28, 31, 72, 100, 106, 113, 116
Collington, 206
Colwall, 106, 107, 109, 199
Colwall Fault, 70
Common Hill, 30, 192
Coneygree Hill, 107
conglomerate, 25, 46, 54, 59, 77, 82, 83, 97, 102, 116, 131, 133, 138, 139, 154, 163, 195, 200
continental drift, 36-41, 43-46, 59
Coppet Hill, 23-27, 31, 139, 154, 155, 169, 170
coral, 44, 53, 92, 103, 105, 106, 107, 109, 111, 113, 115
cornstone, 100, 130, 131, 136, 204
cornstone hills, 3, 15, 31, 132
Covenhope, 118, 119
Cowleigh Park Formation, 97, 102
Cradley, 201
Cradley Brook, 174, 176
Crease Limestone Formation. See Gully Oolite Formation
Credenhill, 3, 12, 21, 130, 132, 165
crinoid, 92, 103, 106, 107, 112, 113, 118
Croft Ambrey, 12, 18-23, 27, 108, 180
Croft Castle, 18, 180
Cromhall Sandstone Formation, 151, 152
Cusop Dingle, 131

Devensian, 127, 132, 173, 176, 177, 186, 187
Devonian Period, 35, 44, 45, 59, 124, 125, 128, 135, 139, 132-41, 143
Dinmore Hill, *viii*, 3, 12, 18, 24, 175, 178
dip, 102, 104, 105
dolomite, 150
Dolyhir Limestone, 105
Downton Castle Sandstone, 100, 101, 120, 121, 127, 134, 198
Downton Gorge, 107, 120, 182, 183, 196, 203
Downton Syncline, 116
Drybrook Sandstone Formation. See Cromhall Sandstone

earthquake, 37, 41, 114, 207
Eastnor, 102, 106, 107, 113
energy, 205

erosion, 6, 10, 28, 33, 44, 48, 52, 58, 61, 62, 102, 115, 118, 130, 132, 134, 140, 154, 165, 166, 167, 199
Escley Brook, 172
eurypterid, 118, 120, 121, 134, 135
extinction, 143

fault, 51, 102, 104, 116
Ffynnon Limestone, 136, 185, 204
fish, 6, 118, 119, 120, 126, 130, 131, 133, 134, 135
Folly Sandstone, 97, 102, 103
Fownhope, 31, 142, 192, 206
Freshwater West Formation, 130, 132, 134, 135, 136, 141, 143, 165, 166, 172, 185, 186, 192, 198, 200

Garnons Hill, 130, 132, 165, 183
Garren Brook, 189
Garway Hill, 139
gastropod, 107, 111, 120, 135
glaciation, 15, 31, 52, 53, 81, 89, 94, 132, 176-85
gold, 197
Golden Valley, 132, 172, 198, 202, 205
Gondwana, 43-45, 57, 59, 81, 93, 94
Goodrich, 169
Goodrich Castle, 23, 26, 139, 144, 200, 201
Gorsley, 113, 116, 119, 121
Gorsley Anticline, 96
Gorsty Knoll, 99
graptolite, 53, 87, 89, 92, 113, 114, 115, 116, 118, 124, 126, 127, 138
gravel, 174, 178, 182, 188
gravel pit
  Bodenham, 193, 201
  Lugg valley, 178
  Mathon, 174, 201
  Stretton Sugwas, 201
  Wellington, 201
  Yatton, 183
Great Doward, 24, 137, 147, 170, 171, 206
Gully Oolite Formation, 148, 149, 150, 151, 171
Gurney's quarry, 113

Hackley Limestone, 133
Haffield Breccia, 61, 159, 160, 161
Hanter Hill, 14, 17, 77, 78, 176, 177
Haugh Wood, 28, 30

Haugh Wood Formation, 103
Hay-on-Wye, 2, 15, 131, 168, 178, 179, 189
Hell Wood channel, 181
Helsby Sandstone Formation, 142, 163
Hereford, 2, 31, 175, 178, 197
    Building stones, 5, 141-42, 198
    Cattle, 205
    earthquakes, 207
Herefordshire Beacon, 7-11, 61, 67, 70-72, 106, 107
Hergest Ridge, 12, 16, 17, 19, 78, 97, 113, 114, 118, 166, 176, 180, 181
High Wood, 102
hillfort. See archaeology
Holly Brook, 176
Hollybush Sandstone, 84, 85, 89
Holme Lacy, 123, 125
Howle Hill, 54, 139, 147, 148, 152, 155, 190, 191, 196
Humber Brook, 174, 176, 177
Huntley Hill Formation, 97, 102, 103
Huntsham Hill, 23, 139, 140, 141, 147, 186
Huntsham Hill Conglomerate, 23, 139, 140, 139-41, 151, 155, 165, 190, 197

Iapetus Ocean, 93, 94
igneous rocks, 40, 46-47, 62-65, 68, 72, 75, 77, 78, 83, 103, 166
industry, 195-8

kettle Hole, 180
Kington, 2, 11, 12, 16, 17, 101, 102, 113, 116, 118, 120, 121, 177, 178, 204
    Building stone, 198
    Moraine, 180
    Radon, 207
Kinsham, 184, 185, 189

lag deposit, 119, 120
landslip, 167, 185-86
    Black Mountains, 185
    Dudale's Hope Valley, 186
    Huntsham Hill, 140, 186
    Shortwood Farm, 186, 207
    Woolhope Dome, 28, 96, 186, 206
Ledbury, 2, 9, 10, 13, 31, 97, 101, 106, 107, 113, 116
Ledbury Hills, 10, 61, 71, 95
Ledwyche Brook, 178

Leinthall Earls, 116
Leinthall Earls Fault, 19, 95, 107, 108, 119, 156, 184, 187
Leinthall Starkes, 19, 108
Leintwardine, 20, 21, 22, 116, 118, 188
Leominster, 2, 142, 177, 187, 197
Letton Lakes, 15, 16
Lewis, TT. See Murchison, RI
Lingula, 120
Little Cowarne, 134
Little Doward, 24, 137, 147, 169, 170, 171
Little Hill, 107
Llandovery Series, 100, 101
Llanelly Formation, 150, 151
Lockster's Pool, 189
Longhope, 109
Longmyndian rocks, 77, 79, 89
Lower Dolomite Formation. See Barry Harbour Limestone
Lower Limestone Shale. See Avon Group
Lower Ludlow Siltstone, 28, 100, 113, 118, 119
Lucton, 22, 180
Ludlow, 18, 20, 99, 101, 120
    Anticline, 19, 23, 27, 96, 106, 107, 116, 182
        See also Wigmore Dome
    Research Group, 99
Ludlow Bone Bed, 120
Ludlow Series rocks, 22, 93, 99, 100, 114-19

M50 motorway, 103
magnetism, 36, 39, 40, 41, 125, 126, 161
Malvern Fault, 11, 55, 57, 58, 60, 61, 71, 79, 82, 89, 95, 97, 116, 153, 155, 156, 157, 158, 159, 163
Malvern Hills, 7-11, 55-77, 153-58
    Evolution, 57-62, 153-58
    Railway tunnel, 70
    Rocks, 62-67
    Scenery, 7-11
    Sites, 67-77
Malvern Quartzite, 59, 77, 79, 82, 83
Malverns Complex, 40, 56, 58, 62-75, 75
map, geological, 4, 29, 95, 153
Marcle Hill, 31, 105
Martley Rock, 79-80
Mathon, 133, 174, 176, 199
Mathon River, 174, 176, 177, 183, 195, 201
May Hill, 10, 96, 97, 100, 102, 103, 107, 116, 121, 155, 157, 158

May Hill Sandstone, 100, 101
Merbach Hill, 131, 202
Mere Hill Wood, 118
Mesozoic Era, 162–63
metamorphic rocks, 47-48, 54, 58, 62, 66, 67, 68, 69, 70, 72, 74, 75, 87, 88, 128, 132, 174
Midland Platform, 82, 94
Midsummer Hill, 21, 67, 73, 74
Miss Phillips Conglomerate, 103
Moccas, 142, 180, 193, 199, 202
Moor Cliffs Formation, 79, 101, 120, 121, 123, 125, 127, 129, 130, 131, 132, 156, 165, 167, 177, 192
Mordiford, 3, 31, 175, 192, 193
Mortimer Forest, 19, 99, 107, 113, 121
Much Marcle, 116
Much Wenlock Limestone, 20, 27, 28, 31, 72, 100, 106, 109, 111, 112, 106-13, 116
Murchison, RI, 1, 11, 16-17, 23, 79, 91, 99, 100, 102, 103, 197
   and Lewis, TT, 16, 22
   and Sedgwick, A, 16, 31, 91, 100, 123

Nash Scar Limestone, 102, 103, 105
Nash Wood, 102
nautiloid, 92, 113, 114
Neath Disturbance, 97, 105, 153, 154, 172
Neogene Period, 167-8
News Wood, 76
nodular limestone, 111, 113

Olchon
   Brook, 172
   Valley, 13, 14, 123, 124, 137, 204
Old Radnor, 114, 121
Old Red Sandstone, 5-6, 91, 94, 96, 100, 116, 120, 121, 123-44, 200
Orcop Hill, 139
Ordovician Period, 35, 44, 53, 58, 59, 87-89, 93, 94, 97, 100, 101, 135
Orleton, 178, 180

Palaeogene Period, 167-8
palaeogeography, 43–45
   Cainozoic, 168
   Cambrian, 43, 82
   Carboniferous, 145
   Devonian, 44, 129, 143
   Ordovician, 88

Permian, 45, 159
Silurian, 44
Pangaea, 36, 39, 45, 153, 159
Park Wood, 109
patch reef, 109, 111
peat, 152, 197, 207
Pedwardine, 77, 79, 89
Pembroke Limestone Group, 148, 151, 152
Pennant Group, 152
Pentaloe Brook, 28, 192
Penyard Park, 139, 140, 190, 191, 197
Permian Period, 10, 35, 45, 53, 60, 61, 153, 158–62
Perrystone Hill, 105
Picturesque. See Artistic movements
Pitch Coppice, 99, 121
plate tectonics, 36-42
Pontrilas, 168, 198
Precambrian Period, 4, 7, 15, 36, 43, 53, 55–80, 81, 82, 84, 86, 89, 94, 101, 158
Presteigne, 11, 97, 102, 105, 187, 194
   Basin, 183, 184, 189
   building stone, 102
Pretannia, 93, 94
Přídolí Series rocks, 100, 120-21
Prior's Frome, 28
Psammosteus Limestone. See Chapel Point Limestone

Quarry
   Aymestrey, 117
   Bringewood Forge, 120
   Brockhill, 109
   Capler, 142
   Causeway, 147
   Chase End, 75
   Church Hill, 118
   Clifford's Mesne, 102
   Cradley, 200
   Credenhill, 130
   Dingle, 68
   Dolyhir, 78
   Drybrook, 149
   Eastnor, 113
   Gardiner's, 70, 200
   Gore, 78
   Great Doward, 147
   Gullet, 56, 71, 72, 73, 101, 103, 200
   Gurney's, 113

Haffield House, 159, 160
Harewood End, 138, 139, 198
Hayslad (or Dogleg), 69
Hobbs, 109
Hollybush, 55, 74, 200
Hollybush Middle, 83
Knight's, 105
Leinhall Earls, 20
Leinthall Earls, 19, 108, 200
Linton, 113, 116, 121
Linton Tile Works, 133
Lord's Wood, 150, 151
Lower Earnslaw, 69
Lowe's Hill, Bartestree, 156
Mere Hill Wood, 118, 119
Mocktree, 116, 118
Nash Scar, 102, 103, 105
North, 72
Park Wood, 109
Perton, 116, 117, 121, 186, 200
Pitch Coppice, 99, 113
Pontrilas, 198
Scutterdine, 105
Slasher's, 74
Sleaves Oak, 31
Strinds, 78
Swardon, 28
Sycamore Tree, 103
Tank, 200
the Knob, 112
Upper County, 69
West of England (or County), 69
Westonhill Wood, 132, 133, 198
Whiteleaved Oak, 74, 85, 86
Whitman's Hill, 106, 113
Woolhope, 105
Quartz Conglomerate. See Huntsham Hill Conglomerate
Queenswood, *viii*, 144

radon, 207
Raggedstone Hill, 55, 67, 74, 86
Raglan Mudstone Formation. See Moor Cliffs Formation
Red Daren, 124
Rheic Ocean, 93
Ridgeway, 71, 107, 116
River Arrow, 3, 165, 180, 181, 201, 202
river diversion, 174, 192

River Dore, 172, 205
River Frome, 3, 165, 168, 190, 192, 197, 206
River Leadon, 3, 9, 31
River Lugg, 3, 20, 22, 31, 127, 165, 168, 174, 175, 177, 178, 182, 183–85, 187, 189, 193, 201
    Evolution, 177
    Terraces, 177, 178
River Monnow, 3, 13, 165, 168, 172
River Rea, 178
River Teme, 3, 12, 20, 120, 174, 188, 196, 203
    Evolution, 107, 178, 182, 183, 187
    Terraces, 177, 178
river terrace, 26, 177, 178
River Wye, 3, 12, 23, 27, 91, 123, 147, 165, 168, 175, 187, 189
    Abandoned meander, 139, 190, 191
    Evolution, 26, 163, 167, 183
    Gorge, 24, 137, 148, 165, 166, 169-72, 175, 193, 196, 201, 202, 203
    Terraces, 177, 178
Romanesque. See Artistic movements
Romantic. See Artistic movements
Ross-on-Wye, 2, 12, 154, 203
    Abandoned meander, 190, 191
    Building stone, 26, 138
    Flint tool, 195
rottenstone, 115
Rough Hill Wood, 102
Rushall Formation, 121

Sapey Brook, 175
Seager Hill, 105
Sedgwick, A. See Murchison, RI
Senni Formation, 136, 165, 172, 185
settlement, 2, 21, 139, 201
Severn Vale, 82, 157, 158, 159, 163
Shelveian event, 44, 59, 89
Shobdon, 142, 180, 183
Shobdon Hill, 12, 19, 180, 184
Shucknall Hill, 97, 154, 165, 192
Silurian Period, 44, 59, 87, 89, 91-121, 124, 127, 197
Sitch Wood, 116
Sned Wood, 118, 184, 185
soil, 5, 10, 12, 45, 88, 106, 123, 127, 130, 151, 195, 202, 203, 204, 205
    fossil, 126
St George's Land, 45, 147, 150, 153

233

St Maughans Formation. See Freshwater West Formation
Stanner Rocks, 14, 15, 17, 77, 78
starfish, 110, 118
stratigraphic names, 33, 100, 101, 126, 146
Staunton-on-Arrow, 202
Staunton on Wye, 16, 179, 183
Stoke Prior, 176, 177, 178, 198
Storridge, 105, 106, 107, 113
stratotype, 99, 121
Stretford Brook, 176, 177, 178
Stretton Sugwas, 179, 183, 201
strike, 104, 116
stromatolite, 109, 111, 150
stromatoporoid, 92, 107, 109
submarine canyon, 93, 116, 118
Suckley, 107
Swansea Valley Disturbance, 153, 156, 168, 172, 189
Swinyard Hill, 72, 73, 74, 101
Symonds Yat, 24, 147, 149, 152, 164, 169, 176, 186, 196, 206

Tan Brook, 192
Temeside Mudstone, 101, 127
The Beck, 102
The Knob, 105
The Leasows, 178, 179
time, geological, 35, 87, 93, 98, 167, 173
Tintern Sandstone, 140, 155, 190, 197
topography, 2-3, 32, 104-8, 153-55, 165-94
Tower Hill, 105
Townsend Tuff, 131
trace fossil, 83, 129
    Beaconites, 133
transgression, marine, 97
transport, 3, 205
Trenchard Formation, 152, 155
Triassic Period, 162–63
trilobite, 35, 53, 81, 86, 87, 92, 106, 107, 110, 113, 115, 118
tufa, 143, 193, 199

unconformity, 50, 51, 84, 85, 103, 139, 140, 146, 152, 158
Uniformitarianism, 94, 115
Upper Coal Measures. See Warwickshire Group
Upper Ludlow Siltstone, 28, 100, 115, 116, 117, 119

Variscan orogeny, 45, 59, 97, 151, 153-57, 161

Wall Hills, 31
Wapley Hill, 12, 19
Warren House Formation, 66, 73, 75-77, 82
Warwickshire Group, 152
Wellington, 201
Welsh Basin, 82, 88, 94, 97, 114, 116
Wenlock Series rocks, 99, 100, 103-14
Wenlock Shale. See Coalbrookdale Formation
West Malvern, 102
Westonhill Wood, 128, 130, 132, 133
Whitchurch, 201
Whitehead Limestone Formation. See Llanelly Formation
Whiteleaved Oak, 86, 89, 159
Whiteleaved Oak Shale, 86, 87
Wigmore, 22
    Castle, 182, 202
    Dome, 96, 106, 107, 121, 156, See also Ludlow Anticline
    Glacial lake, 20, 108, 182, 183, 188
    Vale of, 18, 20, 23, 44, 182
Wigpool, 139, 141, 152, 155, 164
Woolhope, 3, 27, 31, 44, 96, 100, 103, 105, 112, 116
Woolhope Club, xi, 6, 10, 11, 14, 22, 29, 31, 76, 79, 80, 135
Woolhope Cockshoot, 28
Woolhope Dome, 3, 10, 24, 28, 30, 27-32, 61, 96, 97, 105, 106, 107, 116, 117, 121, 154, 156, 165, 186, 192, 198, 200, 204, 206
    Map, 29
Woolhope Fault, 105, 116, 153, 156, 157, 159
Woolhope Limestone, 27, 31, 72, 100, 102, 103, 105, 106, 116
Woolhope Naturalists' Field Club. See Woolhope Club
Worcester graben, 60, 61, 62, 163
Worcestershire Beacon, 7, 10, 11
Worsell Wood, 14, 17, 77, 78
Wyche Formation, 103
Wye glacier, 15, 20, 178, 179, 180, 182, 183, 201

Yartleton Formation, 103
Yazor Brook, 183

Also from Logaston Press: www.logastonpress.co.uk

The Story of Hereford
Edited by Andy Johnson & Ron Shoesmith

This book tells the story of Hereford in breadth and depth, and includes the results of recent research and archaeological investigation. Alongside more familiar aspects of the city's history – how it fared in the Civil War, the foundation and history of the cathedral, the navigation of the Wye – there is new material on Saxon Hereford, medieval trade, Georgian Hereford and the activities of freehold land societies in the Victorian period. There is also information on less well known aspects of the city's past, including Hereford's prominence as a great centre of scientific and other learning at the end of the 12th century, and the use of the city as a base by Simon de Montfort, and also by Prince Henry in the wars with Owain Glyn Dwr. Whether you are familiar with Hereford's history or completely new to it, there is much here to interest, intrigue and surprise.

Paperback, 336 pages with over 160 colour and 50 mono illustrations   Price £15

Walking the Old Ways of Herefordshire:
the history in the landscape explored through 52 circular walks
by Andy and Karen Johnson

Each walk passes or visits a number of features about which some background information is given. These include churches, castle sites, deserted medieval villages, landscaping activity, quarrying, battle sites, dovecotes, hillforts, Iron Age farmsteads, Saxon dykes and ditches, individual farms and buildings, squatter settlements, almshouses, sculpture, burial sites, canals, disused railway lines – to name but a few, and including some that can only be reached on foot. The walks have also been chosen to help you explore Herefordshire from south to north, west to east, from quiet river valleys to airy hilltops, from ancient woodland to meadows and fields, from remote moorland to the historic streets of the county's towns, and of course Hereford itself. The walks range from 2½ to 9½ miles in length. The combination of photographs and historical information make this more than simply a book of walks, but also a companion to and celebration of Herefordshire.

Paperback, 384 pages, over 450 colour photographs and 53 maps   Price £12.95

The Archaeology of Herefordshire: An Exploration
by Keith Ray

Keith Ray was Herefordshire's County Archaeologist between 1998 and 2014, during which time he generated a wide range of exploratory projects, including many excavations. Much new knowledge and understanding of Herefordshire's archaeology has been gained as a result. In this study, he has described what is now known of the county's archaeology, assessing both the work of past generations and the discoveries of this modern era of enquiry. New insights are gained on the activities and rituals of our Neolithic ancestors, the ebb and flow of beliefs at the transition of the Neolithic into the Bronze Age, the shape of Saxon Hereford, the extent of an iron-working industry showing that Herefordshire was once a surprisingly industrial county, and much besides.

Paperback, 448 pages with 230 colour illustrations   Price £15

Also from Logaston Press: www.logastonpress.co.uk

The Hidden History of Ewyas Lacy in Herefordshire
by Priscilla Flower-Smith

The 'hidden history' relates to the details of the lives of the residents of Ewyas Lacy that lay buried in the wills and inventories they left in the three hundred years from the mid 1500s. We see the rise and fall of fashions, from the clothes worn and the furnishings coveted and treasured to the crops grown and stock kept. The local mercer emerges as an important figure. Many women are found to have run farms, and some widows became powerful in their own right. There also emerge hints as to how sin was dealt with by the church and changing attitudes to religion. Intertwined throughout are many family dramas, both comic and tragic, played out through the surprisingly eloquent pages of legal documents. While the book focuses on the period 1550-1850, the initial chapters provide an overview of the area's earlier history and the conclusion brings the story up to the present time.

Paperback, 224 pages, 37 colour and 28 mono photographs and maps   Price £12.95

On the Trail of the Mortimers
With a Quiz and an I-Spy competition
by Philip Hume

This book both gives a history of the Mortimers (notably in their actions and impact on the central Marches) and includes a tour that explores the surviving physical remains that relate to the family. Partly through the good fortune of having an unbroken male succession for over 350 years, and also through conquest, marriage and royal favour, the Mortimers amassed a great empire of estates in England, Wales and Ireland; played key roles in the changing balance of power between the monarchy and nobles; deposed a king and virtually ruled the kingdom for three years; became, in later generations, close heirs to the throne through marriage; and seized the throne through battle when a Mortimer grandson became King Edward IV. A Quiz and an I-Spy have been designed to give pleasure to families wishing to find out more, with the successful completion of the latter leading to a certificate issued by the Mortimer History Society.

Paperback, 144 pages with over 75 colour photographs, maps and family trees   Price £7.50

The Parish that Disappeared; a History of St John's, Hereford
by Liz Pitman

From early in the 12th century until its final dissolution in 2012, the parish of St John's was at the heart of Hereford. Its houses and shops clustered around the cathedral, but it also encompassed land in Hereford. The relationship between the cathedral and the parish varied between amity and tension. But the history of the parish is as much the story of its characters, both the clergy who served it and the parishioners who lived within its bounds. There included paupers, old sailors, a comedian, actors, feltmakers, wool staplers, Italian apprentices and whores, whilst body snatchers also make an appearance.

Paperback, 128 pages with 30 colour and 30 mono illustrations   Price £10